D1554365

The Bell System and Regional Business

The Johns Hopkins / AT&T Series in Telephone History
Louis Galambos,
Series Editor

The Telephone Enterprise: The Evolution of the Bell System's Horizontal Structure, 1876–1909, by Robert W. Garnet

The Anatomy of a Business Strategy: Bell, Western Electric, and the Origins of the American Telephone Industry, by George David Smith

From Invention to Innovation: Long-Distance Telephone Transmission at the Turn of the Century, by Neil H. Wasserman

The Bell System and Regional Business: The Telephone in the South, 1877–1920, by Kenneth Lipartito

The Bell System and Regional Business

The Telephone in the South, 1877–1920

Kenneth Lipartito

The Johns Hopkins University Press
Baltimore and London

Printed in the United States of America
The Johns Hopkins University Press, 701 West 40th Street, Baltimore, Maryland 21211
The Johns Hopkins Press Ltd., London

Winner of the 1987 Allen Nevins Prize in History.
Published by special arrangement with the Columbia University Press.

Portions of chapters 5, 6, and 8 appeared in the *Journal of Economic History* (June 1989).

The paper used in this publication meets the minimum requirements of American National Standard for Information Sciences—Permanence of Paper for Printed Library Materials, ANSI Z39.48–1984.

Library of Congress Cataloging-in-Publication Data

Lipartito, Kenneth, 1957–
The Bell system and regional business : The telephone in the South, 1877–1920 / Kenneth Lipartito.
p. cm. — (The Johns Hopkins/AT&T series in telephone history)
Bibliography: p.
Includes index.
ISBN 0–8018–3797–9 (alk. paper)
1. Southern Bell Telephone and Telegraph Company—History.
2. Telephone companies—Southern States—History. 3. Telephone—Southern States—History. I. Title. II. Series.
HE8846.S55L57 1989
384.6'06'575—dc20 89–32037
CIP

To my mother and to the memory of my father

Contents

Editor's Introduction

In 1967 Paramount Pictures released *The President's Analyst,* a comedy starring James Coburn as the ill-starred, White House psychiatrist. Following a standard 1960s format, the movie landed barbs in a number of formidable institutions, including the Central Intelligence Agency, the Federal Bureau of Investigation, and the telephone company. The Bell system was portrayed (under a different name, of course) as a ubiquitous, omnipotent, high-tech organization which, unlike the FBI, wanted to be liked. Even though it employed robots (that is, managers with "Bell-shaped heads"), they were bland fellows with pleasant voices. Perhaps that was why in the film's humorous ending, the hero celebrated his escape—unaware that he, his lover, and his secret-agent friends were all the while being watched by the phone company's concealed TV camera. Rumor has it that Bell executives who were shown the film at management seminars wildly cheered the conclusion.

While Kenneth Lipartito's book is a serious scholarly monograph and not a sixties spoof, the author actually touches on a number of the themes raised by the movie. As Lipartito explains, we frequently view the triumph of large-scale systems as an inevitable, almost mechanistic, and unequivocally beneficial aspect of modern society. These systems are, after all, the carriers of modern technology. Because they are the agents of technological change, their opponents are by definition the enemies of efficiency and progress. Not so, Lipartito declares. Modern large-scale systems bring certain specific advantages, but their costs in particular settings outweigh their benefits. We need in each case to deconstruct the process of system building, analyze the alternative paths, and show how the costs and benefits shifted over time for different groups and regions within the larger society. We need to know how the system builders deployed their power and resources and how they made their path to the future appear to be the only rational course for society to follow. We need as well to understand from several perspectives the conflicts system building engendered.

This is exactly what Lipartito does with great skill and sensitivity in his history of the introduction of the telephone in the southern United States between 1877 and 1920. Initially, the South was short on entrepreneurs who could muster the technical skills, capital, and managerial ability needed to launch successful local phone companies. Regional patterns of entrepreneurship in New England and the Midwest were better suited to development than those in the South. Thus imported talent took the helm. "Cross-cultural entrepreneurs" such as James Ormes and E. J. Hall led the process of system building until a cadre of indigenous businessmen emerged to fill the top slots at Southern Bell. One can see very clearly in this story how intertwined and important in development were a region's supply of potential leaders, its sources of capital, and its array of supporting business institutions.

In the course of describing these events, the author advances our knowledge of entrepreneurship in several interesting ways. He shows that there were various entrepreneurial styles, rooted in different economic settings and likely to produce different economic outcomes. Local businessmen who opposed Bell system innovations were thus not necessarily the opponents of progress. Instead, they had a different entrepreneurial vision. They looked to a mixed portfolio of local investments to produce favorable results for themselves, their locality, and their region. Ormes and Hall, by contrast, were spokesmen for the Bell style of innovation, which embraced a relatively expensive, standardized technology and stressed the virtues of being part of an all-embracing, national telephone network. Where these two visions clashed—in the South, in particular, but also in the Midwest—the opponents used all of the political as well as economic weapons they had to win control of the market. The results were not foreordained by Bell's technological superiority. Lipartito clearly establishes the problematical nature of this struggle for the control of telephony. He is evenhanded in his evaluation of both Gulliver and the Lilliputians: AT&T "cajoled, bribed, colluded, and at times imposed its design on towns and cities." But this did not make the firm "particularly reprehensible; its competitors engaged in similar tactics."

Neither side in the protracted war made any distinction between political and economic battlefields. All of the combatants used municipal and state governments in an effort to strengthen their positions. While Lipartito finds examples of public institutions that favored one side or the other, he does not see evidence of conspiracy or capture of the regulatory process. Rather, he finds that some public officials were simply better placed and more effective in achieving ends that promised the most good for the most people in their state or locality. In this and other evaluations, Lipartito uses a comparative regional analysis that helps us see exactly how poor southern political leadership was and what that cost the region.

Throughout its early history, the expanding Bell system engendered conflict in every part of the nation, but the struggles were especially fierce in the South. There, municipal governments wielded their powers over franchises and rights-of-way in an effort to control telephony. Both the system and the region remained in flux for several decades. In the South the transition from an older set of values and institutions was unusually painful. After the Bell patents expired in 1894, independent competitors in the South and elsewhere rushed in, first to fill the gaps in rural and small-town markets and then to challenge Bell in the cities and even in toll and long-distance service. Lipartito gives these contestants names, faces, and a sympathetic hearing. We learn about Pan Electric, the Alabama Telephone and Construction Company, and the Interstate Telephone Company of Durham, North Carolina. We thus understand the human as well as the organizational and economic dimensions of the dual structure—evenly divided between Bell companies and independents—the industry had acquired by 1907. We also can see why it was so astute for the Bell system, led now by the redoubtable Theodore N. Vail, to embrace state-level, rate-of-return regulation and to launch an aggressive policy of interconnection with heretofore independent companies.

While during the late nineteenth century, the telephone in the South lagged behind the Northeast (in technology as well as extent of service), two formidable innovations emerged from Dixie.

One was a pattern for the 1879 settlement between Western Union and the Bell interests, a contract that secured the patent monopoly until 1894. The other was the policy of interconnection with independents, a program that evolved in the South and became a national policy of the system after 1907. These two turning points were decisive in shaping the development of Bell's national network.

In the end, of course, the network, the high-tech system, the modern bureaucratic organization triumphed over localism and regionalism (as it did over heroic individuals in *The President's Analyst*). Lipartito describes and dissects this outcome with the same care that he displays in treating the regional opponents of the system. As he convincingly argues, the blend of Bell's technologically standardized and centralized network with a decentralized structure of operating companies under state regulation helped nudge the South toward full participation in America's twentieth-century industrial economy. When an entrepreneurial style, a technological style, and a political style meshed as well as they did with the Bell system by 1920, you had a very stable institutional setting in the South and the rest of the nation. Indeed, this combination would remain intact until the crisis that led to the fall of the Bell system in the 1980s.

In the long run, Lipartito concludes, most of the region's people benefited from this transformation. The Bell system "offered much that the people of the region lacked—rapid communications, contact with the outside world, large-scale technologically sophisticated industry." By blending a measured analysis of the results with a penetrating, comparative study of the social, economic, and political processes that accompanied the emergence of America's telephone network, Lipartito has written a volume that we are especially pleased to have in the Johns Hopkins/AT&T Series in Telephone History.

Louis Galambos

The AT&T Corporate Archives

It is the policy of AT&T to preserve all company records and artifacts that document its history:—the evolution of its structure and organization, the development of its products and services, and the evolution of corporate policies. Therefore, in addition to record retention requirements imposed by law and retention standards adopted by individual departments for their own purposes, the potential corporate historical value of records will be taken into account in the retention and disposal process. The company's central archival organization has responsibility for assessment of the historical importance of all company records and artifacts and the designation, accession, and preservation of archival records.

The company policy is to stimulate scholarly awareness and use of the materials. AT&T has undertaken to make these records and artifacts available for corporate and approved scholarly reference and use through a systematic program of accessioning and cataloguing, by the establishment of adequate reading room facilities, and by direct preparation of publications.

Acknowledgments

In a project that has gone on now almost six years, one is bound to incur many debts, more than can be properly acknowledged. I would like to at least try to thank some of those who made this book possible.

Most of my research was carried out at the AT&T archives in New York. There, Robert Lewis and Robert Garnet provided invaluable assistance in research and advice on the development of this project. Through Robert Lewis, AT&T generously provided financial support in 1985 that helped me to complete my work on schedule. Mildred Daghli devoted innumerable hours of her time to helping me find and retrieve records and generally brightening my days of research. I benefited greatly from the help of Alan Gardner, whose stimulating discussions about the history of telephony I relied on to focus and sharpen my own ideas.

The Southern Bell Telephone and Telegraph Company also provided valuable material, which greatly enhanced the content of this book. Lynn Jones, Anita Smith, and Ralph Payne all aided me in my search for historical sources.

I would like to acknowledge my gratitude to the Georgia State Archives and the Atlanta Historical Society. At an early stage of my research I also benefited from the excellent staffs at the University of North Carolina and Duke University libraries and the North Carolina State Archives.

When this work was still a dissertation at the Johns Hopkins University, I received support, suggestions, and useful criticism from members of that school's history department. Surely my biggest debt in this regard is to Louis Galambos, who suggested that I work on the topic and has lent his many talents to assuring, first, that my work led to a finished product, and second, that it was worth the effort. Lou has continued to provide guidance and counsel through the process of revision. Incisive readings by William Freehling, Carl Christ, Bill Leslie, and Steven Hanke, members of my dissertation committee, also assisted me in my revisions.

Many others have given their time and help by listening to my often incomprehensible ideas or ploughing through early drafts of the manuscript. A careful reading by Naomi Lamoreaux saved me from errors of logic and reasoning, while discussions with Joseph Pratt, Claude Fisher, Marvin Sirbu, and David Gabel helped me to refine and polish my ideas.

Financial support from the Economic History Association, in the form of an Arthur H. Cole Grant-in-Aid assisted me in turning my dissertation into this book.

Finally, two special thanks: One to those friends who lived through this project with me, as sometimes unasked but always gracious hosts to my obsession with the telephone industry; and one to Horace—a cat who truly knows where he's at.

The Bell System and Regional Business

Introduction

Of all the achievements of American business in the nineteenth century, perhaps none was more impressive than the building of the great systems in transportation, communications, and electric power. Spanning the continent, their size, cost, and technical wizardry captured the imagination of supporters and critics of big business alike, suggesting at once the limitless size and scope of corporate organization and the complete helplessness of preexisting ways in the face of technological progress. As the infrastructure that made continuing business evolution possible, these systems were the cutting edge of the managerial revolution in American business that produced large-scale firms in other key sectors of the economy. As the most far-reaching of new business organizations, they penetrated nearly every island community in the United States. To paraphrase one historian of this period, they helped to bring about the incorporation of American society.[1]

Both to those who observed the building of these systems with awe and to those who followed it with anxiety, one lesson seemed clear: Nothing could stop the further drawing together of the nation into a single market and a single culture. It was simply a matter of time before railroad, telegraph, telephone, and electric lines bridged the gaps separating the cities, towns, and rural communities of the nation.[2] No commitment to traditional ways could slow these overwhelming forces, it was believed; America as a society was wholly committed to technological and economic modernization, no matter how disturbing a form it might take.

In retrospect, this common wisdom seems suspect. Historians know that change is never as dramatic as those in the midst of it believe. Both the causes and consequences of change often appear different at a distance than they do close up. Studies of modernization and the responses to it in nineteenth-century America have unearthed a trove of protest, both public and private. Populist farmers fought with railroad system builders; urban politicians sought to regulate the private parties in control of public utilities. If these dissenters made up only a small percentage of the public, they nonetheless raised questions and voiced anxieties that were felt by others. No matter how inexorable the growth of large-scale systems may have seemed, some people at least tried to mark out different paths of history.

Even where political resistance was weak, the growth of the new systems did not come about overnight. Simply constructing a system that had to be physically connected and comparable in all places proved a monumental task. The first system-builders faced a world filled with regional variations and local differences.[3] Away from a few large urban places, the resources necessary for the completion of an integrated system were hard to come by and demand conditions proved unpredictable and quixotic. Achieving standardization and coordinating functions in such an environment involved not only technical competence, but feats of entrepreneurial acumen and organizational creativity.

Recognizing the complex problems faced by the first system-builders makes the completion of America's transportation, power, and communications networks all the more remarkable. They were not the product of passionless progress, but the dramatic achievements of their architects in the face of adverse conditions. They did not proceed with smooth inevitability but rose over protests and conflicts. Not enough is known about the men and the organizations responsible for these achievements, particularly outside of railroading.[4] How did they overcome political protest and deal with regional variations? What roles did private initiative, public power, entrepreneurial skill, and technological innovation play? And most important, how did the great systems incorporate the nation's isolated island communities with their distinctive cultures into a national matrix linked by physical, financial, and organizational lines?

This book attempts to answer these questions through a case study of one of these systems—the telephone network. It will examine the process of regional system growth by focusing on the coming of the telephone to the South, one of the nation's most backward regions and therefore one of the most difficult to penetrate. Through comparisons with other regions, primarily the Midwest and New England, the study will attempt to isolate those factors most important in the building of the southern network, in order to see just how a national system evolved in the face of adverse regional conditions. Such comparisons will also permit an evaluation of the impact of the telephone system on southern development.

Because of the way the telecommunications network emerged in America, it offers an excellent opportunity to examine the relationship between a large-scale system and a regional culture. In the years immediately following Alexander Graham Bell's invention of 1876, there was no national telephone organization, only a number of decentralized regional operations. Each section of the nation thus responded to the new technology of communications in its own way. Gradually, regional telecommunications came under the control of larger regional companies. Each of these companies operated under a license from the Bell Telephone Company of Boston, which held the only legitimate patents on the basic technology of telephony. Still, throughout the first two decades of the industry, conditions varied widely between places and interregional communications were limited. Indeed, after the original Bell patents expired in 1894, the industry became even more decentralized, as competitors entered the market in great numbers. By 1907, however, this situation had changed dramatically. Through AT&T, the Bell long-distance company, the system-builders had gained control of a significant portion of the telephone industry and had laid the foundations for a national interconnected network of communications. The story of the telephone system is thus one of gradual integration and centralization in the face of strong forces pushing for decentralization and regional variation.[5]

Examining the relationship between systems and regions offers an opportunity to fill a gap in the literature of modern business. Written mainly from the perspective of the firm or industry, business history has given short shrift to the important economic, social, political, and cultural impact big business has had on different parts of the nation. Though it is well understood that the great technological systems of the nineteenth century provided the necessary preconditions for the growth of large-scale industrial firms, less well known is how these systems profoundly altered the economies of regions like the South.[6] Often the first example of modern organization to reach isolated localities in backward sections of the nation, they centralized operations, took coordination and control of vital economic activity out of local hands, and brought to (or imposed on) such places the methods, standards, values, and concepts of more advanced areas. Their potential for inducing deep and pervasive change in regions was thus enormous.

Viewing system development from the perspective of the region also helps to clarify an important but not often noted facet of the emergence of big business. The growth of large-scale organizations brought about a profound change in the relationship between business and its regional environment. This change lay behind much of the protest leveled at big business between 1880 and 1920.

Before emergence of national-scale firms, business in America had been primarily small scale and local. Whether in mining, manufacturing, agriculture, or trade, most business operations were carried on in a restricted local economy. Interregional trade involved not giant wholesalers or multiunit retailers, but numerous individual firms coordinated by the invisible hand of the market. Industrial production was even more locally bound, particularly before the availability of cheap overland transportation.[7] Even merchant houses that participated in international trade carried on what was essentially a local enterprise, selling a product produced nearby, investing their surplus funds in land or other ventures close to home. The vast majority of American businessmen who were not directly involved in the international economy, including the ubiquitous commercial farmer, were even less cosmopolitan in their operations than merchants.

In this world, businessmen of necessity tied their fortunes closely to that of the place in which they lived. As the community grew, so did their enterprises. Demand for their products increased when more people and trade came to town. Investments in land made by the prosperous merchant or manufacturer increased his ties to his community, for land values too could be expected to rise with the general progress of a place. Self-interest dictated that the businessman become a booster for his community. Localism assured that individual business growth corresponded to community development, providing a legitimate basis for business in American life.

It was this web of interests that bound the businessman to the place in which he lived that the modern corporation disrupted. Unlike small businesses, large-scale operations by definition went beyond the confines of any one place. The orientation of the manager of a rail system or the executive of major oil company was not his nearby surroundings, but the national or world economy. Profits were not linked to the progress of a single community, but to the overall efficiency of an entire organization. Managers of such organizations tied their lives and careers to them, moving as they moved, adopting the corporation's point of view and values for their own much as the earlier businessmen tied theirs to their communities. The scale and scope of such operations, moreover, exceeded the ability of any nineteenth-century community to regulate them. In a society long used to a different set of relationships, the tendency of modern business to sever the connection between the firm and its location raised disturbing questions and often excited anxious protests.

Geography, of course, has not been completely eradicated as a factor in the economy even today. Small businesses continue to operate beneath the huge center firms of the economy, much as they did before the 1880s. The owners and operators of small-scale enterprise today exhibit the same local orientation as their forebears, joining rotary clubs, fraternal organizations, and chambers of commerce to promote local progress. Responsible corporations also cultivate ties to the various places in which they operate, and footloose modern managers are still able to set down some roots.

Still, since the rise of big business, the trend has been to make geography less of a factor in business. Multinational corporations have cut their moorings to cities, states, and nations and have extended their reach around the globe. Using sophisticated methods of communication and control, they coordinate their wide array of functions from their central headquarters. Neither geographical nor cultural barriers seem to inhibit them as they pursue new markets, move production off-shore in search of lower labor costs, and seek out favorable political environments in which to operate. In this world, the local economy remains, but only as a shadow of its former self. More and more activities that used to be carried on locally become part of the corporate sector through devices such as the franchise and the dealership. Directly through such institutional arrangements and indirectly, through the importance of center firms to the overall economic health of a nation, local business is dependent on big business.

The global ambitions of corporations today cause Americans to ask what relationship will such enterprises have to the nation as a whole? This question is very similar to the one that people asked in a somewhat different context a century ago, when the first system-builders were molding their enterprises. Then as now, some found reasons to celebrate and some to contest the new direction of business development. What follows is a study of this first era of conflict and change in the relationship between the corporation and its regional environment.

Chapter 1

The Telephone Heads South

In the telephone's first few years not even the most optimistic observer would have imagined that it would become the cornerstone of a giant, nationwide system of communication. Following Alexander Graham Bell's invention of 1876, the telephone industry started out with little capital, a few dedicated pioneers, and a skeptical public. Over time, money men entered the business, taking control from the inventor and his original backers. Years later, other financiers helped to combine all operations into a single, monopolistic firm—AT&T. In the initial years, however, experimentation, decentralization, and regional variation characterized the business. Short of capital, the first Bell company depended on individual agents throughout the nation who dedicated their own resources and talents to promotion for a percentage of the profits of their territory.[1] Town by town, county by county, these men spread the wonders of the talking machine through the United States.

The model for this sort of telephone promotion was provided by New England, the first region of the nation to experience the new device. There the Bell Telephone Company, a patent association for Alexander Graham's invention, licensed its first telephone agents in 1877. Responding to the call, a large number of local businessmen quickly entered the industry and worked to fit the new technology to the demands of the various markets of their region. Operating on their own account as well as the Bell firm's, they drew on local supplies of capital, credit, entrepreneurship, and technical skill as they went about their task.

In cities, early agents built simple but effective telephonic distribution systems.[2] This strategy accorded with the prescriptions of Bell Company executives, who had initially conceived of their product in terms analogous to the telegraph. Before the birth of long-distance telephony, they believed that the telephone would be an adjunct to telegraphic communications. Messages would arrive over telegraph wires and be distributed locally by telephone, in competition with express and message offices. A number of firms in New York and Boston developed this type of service. They cultivated a lively business in these densely populated cities. Customers called the central office, which sent their messages by telephone to local cab companies, livery stables, or the Western Union telegraph office.[3]

While these early message companies worked well in a few large metropolises, customers in less heavily populated places had different needs. Small-town subscribers came from the select few wealthy enough to afford the luxury of telephone service. Before the invention of the telephone switchboard, they did not use central office systems as did large city customers. Instead they relied on point-to-point private lines, which they put up themselves. In smaller towns far from commercial centers such as Boston, bankers ran lines from home to office, merchants from store to depot; manufacturers linked the point of production with the warehouse, and doctors kept in telephonic contact with pharmacists. Individual agents, rather than telephone firms, met this low density demand. Telephone promoters in the New England hinterland often covered an area of several towns or counties in search of business. Though they made additional money by furnishing customers with needed supplementary equipment, such as poles and wire, they faced an uphill struggle. Beyond the city, consumers remained skeptical of the telephone's practical advantages. Responsibility for overcoming these impediments fell to local Bell agents.

In response to these conditions, telephone promotion underwent an important change between 1877 and 1878. The men who had first ventured into telephony were generally substantial businessmen with experience in older forms of communications. In New England, Gardiner Hubbard, trustee for Bell's patents, first appointed C. A. Stearns and J. N. George, Massachusetts telegraph and electrical equipment dealers, as general agents for the region. Well established as representatives for another great communications company, Western Union, armed with a technical understanding of electrical equipment, and well-known to New England's commercial community, they must have seemed ideal choices. In southern New England William Hayward, a New York businessman, served in the same capacity. His qualifications included a wide reputation among businessmen in the area, and capital, which he could commit to the fledgling telephone enterprise.[4]

As promising as these early choices appeared, however, they soon proved inadequate to the task. While general agents such as Stearns and George and Hayward acted in a managerial capacity, overseeing a broad territory, they relied on another group of entrepreneurs to cultivate telephone customers. These men, the field agents or subagents, handled the more mundane chores of telephone promotion in individual towns and cities, changing the reputation of the telephone from that of a toy to a useful means of communications. In New England, field agents were drawn from a fairly large and diverse pool of ambitious applicants.[5] Some were school teachers whose interest in science and technology led them to seize what they hoped was the main chance in an unproven new venture; others were successful local merchants who wanted to add the telephone to their list of subsidiary interests. Most were neither wealthy nor blessed with strong financial resources. New Hampshire agent Wilber Trafton was not considered a worthy financial risk by the R. G. Dun credit reporters. In Maine, R. G. Capin, who simultaneously ran telephone and sewing machine agencies, was forced to take out a personal mortgage to continue his operations. Russell Hyde of Vermont was reported deeply in debt and financially had "never borne a very good reputation."[6] Nonetheless it was through the efforts of men such as these that the telephone began to spread from the major cities of the East Coast to the countryside and the West.

Though they lacked the capital and broad-based contacts of their superiors, subagents brought compensating skills to the new industry. Youth and energy inclined them to give telephone promotion a greater portion of their time than general agents, who often had other well-established business interests and waited to see how the telephone fared in the market before committing to it their resources. Such caution inhibited telephone growth, as Bell Company executives soon discovered. No single formula for successful telephone promotion existed at the initial stages. People were still learning to use the new technology and adapt it to their needs. The limits and utility of telephony remained unclear, as equipment evolved rapidly in the first few years, presenting consumers with new possibilities and new questions. Successful regional development under these conditions depended on agents' entrepreneurial abilities. Local men acting as field agents frequently had the flexibility to deal with this situation, while general agents far away did not. Working in relatively restricted areas, field agents developed a thorough knowledge of their territory and the people in it. With this knowledge they were able to tailor their promotional activities to local demands and keep them sensitive to the vagaries of geography and customer preference.

Still, the problem of capital shortage threatened to overwhelm even the most ambitious field agent. Promoters generally had to show some success before banks and investors would lend them money. Credit problems could therefore be devastating. By 1878, the shortage of funds in the new industry had become desperate. Bell treasurer Thomas Sanders and Boston-based telephone manufacturer Charles Williams wrestled continually over matters of money.[7] Williams demanded payment for the telephones he had supplied, forcing Sanders to dun agents for rental advances. Neither man had the working capital that the telephone needed. Caught in this credit crunch, agents had to find their own sources of funds, or face ruin.

In response to the financial problems of the early telephone industry, field agents drew on the resources of family and friends. Their close contacts with local businessmen, especially manufacturers, also proved valuable, for these men often provided them with the working capital they required. At a time in which the market for industrial securities was in its infancy and banks did not routinely lend money to new ventures, field agents served as finan-

cial entrepreneurs in the fledgling industry. By successfully tapping local resources, they supported the early telephone business in much the fashion Bell executives had hoped.[8]

By 1878, a clear and effective structure had emerged in the New England telephone industry. General agents such as Stearns and George acted in a supervisory fashion, overseeing a broad territory for the Bell Telephone Company, which only rented telephones. Subagents worked in individual towns and did much of the actual work of promotion. These men marketed the new technology and tapped vital supplies of capital for the growing enterprise. In areas close to large cities, general agents and astute promoters founded telephone companies. In the hinterland, point-to-point service, largely built by individual customers with the assistance of Bell agents, sufficed.

While the Bell Company's decentralized procedures worked well in the Northeast, they proved far less effective in the South.* Although the first instruments reached the region by the fall of 1877, telephone development there quickly stalled. Regional figures on telephones are scarce for the early years, but it seems that at a time when almost 10,000 telephones were in use in the nation, the South had at best a little over a hundred.[9] These numbers hardly tell the whole story. For some reason, the quick-witted, technically competent young promoters who became the driving force of the New England telephone industry seem to have been almost wholly absent from the South, or at least not forthcoming in telephony. In addition, the few southern businessmen who did enter the early telephone industry responded slowly to telephone innovations and often proved incapable of overcoming adverse initial conditions.

Initially, the managers of the Bell enterprise treated the South as they did other areas of the nation. In July 1877, Gardiner Hubbard granted to Savannah commission and shipping merchants Richardson and Barnard a large general agency for the deep South.[10] Their territory included most of eastern Alabama and all of Georgia.[11] As with his other appointees, Hubbard expected them

*Here and throughout, South refers to Virginia, North Carolina, South Carolina, Georgia, Florida, and Alabama, roughly the territory served by Southern Bell in this period.

4

to find subagents to spread the telephone to remote parts of the region. In choosing these men, however, Hubbard had tapped the wealth of the old South. Their commission business recalled the days when Savannah and Charleston dominated southern finance. But these cities were no longer the cosmopolitan entrepots of antebellum times. The economy was inland now, in the hands of landowners, country merchants, and builders of the New South. Though the Savannah merchants could contribute money and business expertise to the telephone, they were too far removed from this world to be effective promoters of a new technology. Although their territory included Atlanta and Savannah, they succeeded in renting only 120 telephones by March 1878.[12] Their failure to develop Atlanta particularly angered Theodore Vail, who became Bell's general manager that same year. Vail sensed correctly that the city would become the hub of the southern telephone business. After 1878, the Georgia agents fell even further behind, ignoring important innovations such as the telephone exchange.

One of Hubbard's other early appointees in the South, G. A. Cary, appeared better prepared to meet the demands of the telephone. He was an experienced itinerant insurance salesman from western Alabama. His initial ideas for developing the telephone were more adventuresome than those of his Savannah counterparts. But Cary turned out to be too erratic to be an effective agent. He formulated ambitious plans, such as leasing telephones to mining firms in Mexico and Brazil, stringing a telephone line across Mobile Bay to an island resort, and connecting cotton buyers in Selma and Montgomery by telephone.[13] Yet he conspicuously neglected the burgeoning coal and iron industries of Birmingham, insisting flatly that they would not be interested in telephones.[14] He also remained skeptical about the telephone's general utility in the South, believing it would have only a small market outside a few large cities.[15]

Perhaps the most conspicuous example of failure is the case of C. P. E. Burgwyn, Bell agent for Virginia. Blessed with a lucrative territory which included Richmond, the seaboard South's largest city, Burgwyn seemed well qualified for the new business. He had studied electricity and engineering in the North, and said he had "some slight means" to bring to his agency.[16] Like many of his

northern counterparts, Burgwyn seemed to possess the ingredients for success; he had technical training and was young and ambitious. He even claimed to have some influence with city officials in Richmond and Norfolk. Nonetheless, he failed miserably as a telephone agent.

After his appointment by Gardiner Hubbard in 1877, Burgwyn first tried to interest southern cities in adopting telephones for public use. Contacting the fire alarm superintendent of Richmond and the water works commissioner of Norfolk, he attempted to place telephones in their offices. This same approach had been successful in New England cities. Once prominent public officials took a chance on the new technology, other citizens followed. Burgwyn also mimicked his northern counterparts by soliciting business from local manufacturing firms, such as the Charlottesville Woolen Mills.[17]

Though ambitious, the Virginia agent soon ran into trouble. City superintendents hesitated, claiming that they needed council approval before renting telephones.[18] The wool manufacturers backed down, despite their early interest. A local mining firm, embroiled in a strike, could not experiment with the new technology. The Danville Railroad would have signed a Bell contract, but discovered that its exclusive agreement with Western Union forbade it from using competing communications equipment. Despite Gardiner Hubbard's intervention, they would not budge. Even Burgwyn's original idea of using telephones in divers' helmets at the Norfolk shipyards, a plan heartily endorsed by Thomas Watson, came to nothing.[19] Delays, technical difficulties, and a tepid reception to his suggestion by naval officials made Burgwyn too cautious and deferential. Reluctant to offend those in public office, he never overcame these early problems. By 1878, despite having paid $138.00 in advances and fees to Bell, he had not rented a single telephone.[20]

Discouraged by these early difficulties, Burgwyn began to see his lack of success as the product of misfortune, bad timing, and the "conservatism" of southern businessmen.[21] These beliefs made him even more reluctant to engage in telephone promotion full time.[22] This reaction was understandable, given the problems he confronted, but it did little to advance his situation. Unlike New

England agents, Burgwyn found no substantial merchant or manufacturer to back him financially. As a result, he was hard pressed to meet even the small advances he had to pay. In January 1878, Bell treasurer Thomas Sanders demanded from Burgwyn $48.00 for the telephones he had ordered.[23] Fearing he could not meet these obligations, the agent was forced to reply that he would prefer to pay by check rather than have Sanders draw on his account. Explaining his situation to Sanders some months later, he wrote, "the telephone only forms part of my business, and there are months at a time when I am only in Richmond for a few days."[24]

With Burgwyn's poor showing, Western Union, a new competitor with its own telephone, entered the Virginia market. In other states in the South, Western Union made little headway; but Burgwyn identified the corporation as a major cause of his troubles. City commissioners and railroads "hung back," refusing to rent Bell instruments until they saw which firm triumphed.[25] By the spring of 1878, the Virginia agent was lamenting "the telephone as a telephone is a success, but the undeserving are reaping the rewards."[26]

Burgwyn was not the only Bell agent to feel the heavy hand of competition. Western Union was actually much stronger in New England, as letters from agents there attest.[27] Unlike the southerner, however, northern agents remained optimistic about their chances. They continued to believe that the giant, though powerful, could actually be defeated. As a result, New Englanders responded to competition with more effort, expending more time and money on the telephone. Burgwyn, by contrast, badgered Gardiner Hubbard with letters calling for Bell to fight it out with the telegraph firm in the courts. "They [Western Union] rely on their strength and money to put them through," he complained, and as a result Virginia cities were "all for the Western Union."[28] Burgwyn believed the giant had so "bulldozed" potential customers that Bell's only hope was a victory in a patent infringement suit. Such suits were costly, time consuming, and not always successful, however.[29] Refusing to undertake this risky effort, Bell managers could only offer their southern agent encouragement in his tasks.

Western Union became for Burgwyn the excuse for his failure. He did not heed the parent company's advice, but instead became more cautious. Fearful of committing his time and personal

e 1: The Private Line. Telephone A is connected directly to telephone B. Tele-
es A and B could be independently connected to a third telephone C.

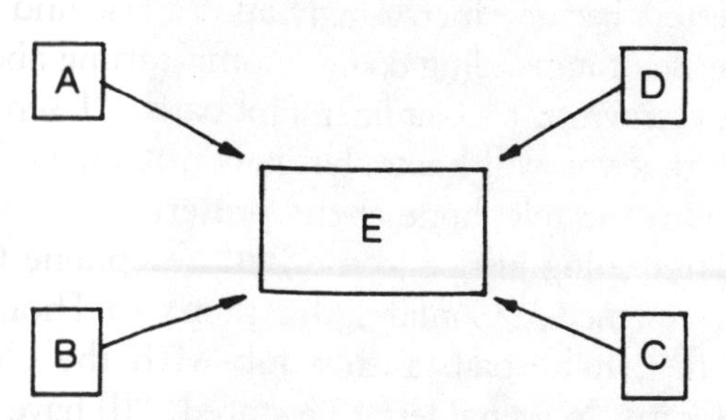

2: District and Fire Alarm Systems. These devices, similar in concept to the ex-
e, did not involve direct interconnections of all subscribers. Instead, callers indepen-
contacted a central office, which relayed their messages to other parties. In the case
lephone fire alarm system, callers could report fires to a centrally located watchman
l boxes around the city.

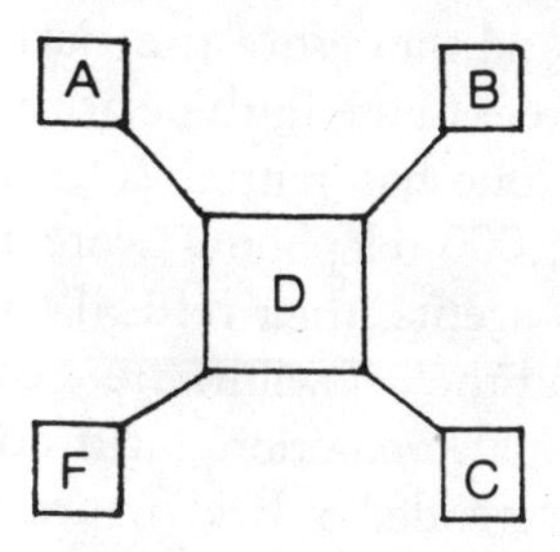

: The Central Office Exchange. Telephones A-D are connected directly to office E
a mechanism (switchboard) that allows each customer to talk with every other cus-

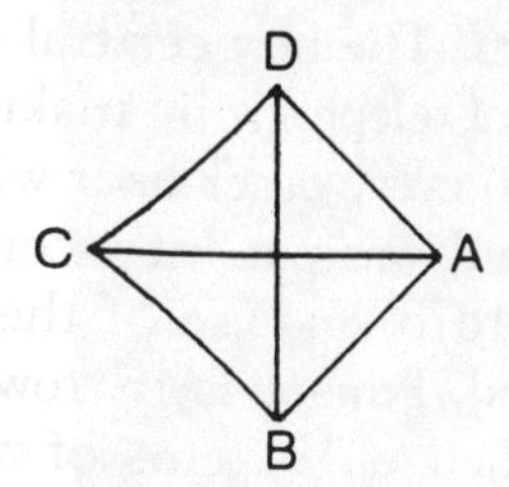

Two-Way Connections. The number of two-way connections needed to link four
rs using private line technology.

means to the new industry, unable to persuade
to rent telephones, short of working capital,
Western Union's competitive price cutting. Tl
tone of his letters became increasingly apolog
projected a sense of impending doom. Compla
Union, Burgwyn wrote to Gardiner Hubbarc
statements so that you will know that it is no
do no more with the telephone at the presen
that he was "not doing justice to the Bell Te
its agent," he resigned.[31] Confiding his plans
Burgwyn wrote that he had taken a job wi
Engineering Corps. In a final letter he stated,
time to push matters as they should be push
think I will leave the prospects of the Bell
favorable light than when I took them up bu
not doing justice to myself or you."[32]

As a result of these disappointing eff
thing, fell further behind the rest of the na
haps 222 telephones were in use in the entir
agency in Maryland alone had rented 280,
United States over 26,000 telephones were
exhibited by southern agents, their refusal
key urban markets, and their unwillingness
capital, organize regional promotion, and
its various demands portended a dark futu
phone industry.

The southern record showed little
the invention of the telephone switchbo
time telephone service had consisted prin
terconnecting two points. The new centr
cally altered the nature of telephony by m
telephone user to talk to every other user
fig. 1–5). The switchboard was put into
New Haven in January 1878, and use of
throughout New England. Promoters in
new technology in "exchanges," systems
to interconnect subscribers.[34] Yet despite
telephonic communications made possi

Figu
pho

Figure
chang
dently
of a te
via ca

Figure
through
tomer.

Figure 4
subscrib

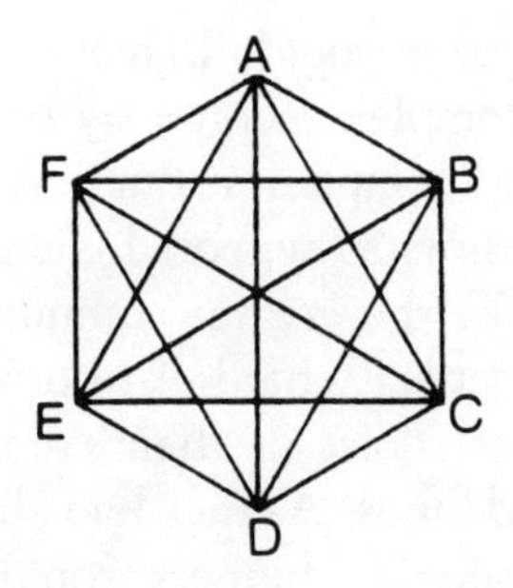

Figure 5: Two-Way Connections. The number of two-way connections needed to link six subscribers using private line technology.
(*Source:* Bell Laboratories, *A History of Engineering and Science in the Bell System* [New York, 1975], vol. 1, 467-72. Kempster B. Miller, *Telephone Theory and Practice* [New York, 1905], 170-75.)

southern entrepreneurs had no more success with it than they had with private lines. So long as the southern telephone industry depended solely on promoters like Burgwyn and Richardson and Barnard, there were no exchanges in the South.

In an 1878 letter to Thomas Sanders, C. P. E. Burgwyn rejected as "uneconomical" the suggestion that he start central office operations in the cities of his territory.[35] Though the Bell treasurer tried to persuade him that he was wrong, the Virginia agent steadfastly clung to his belief. Similarly, Richardson and Barnard approached the exchange with the same caution they had exhibited toward the telephone. Refusing to promote them in their territory, they again cited the different business climate of the South as their reason, stating, "business conditions are so different than in the North [that] but a few private houses would use the system."[36] As this statement suggests, the failure of southern agents to endorse telephone exchanges flowed from the same set of factors that had inhibited them from aggressively promoting telephones. In both cases their evaluations of risk and opportunity were to blame.

The cost of the new technology also dissuaded southern agents from embracing it. Exchanges required greater amounts of capital than did private lines, and expenses which had been paid by customers—poles and wire—now became the responsibility of agents.[37] These costs could quickly surpass the means of even wealthy promoters, if the exchange multiplied telephone demand as Bell executives hoped it would. To meet the added expenses and deal with the more complex technology of the exchange, agents had to form capitalized companies. Burgwyn, unsuccessful in convincing southern investors to support his early ventures, was even more reluctant to make the greater commitment of resources the exchange required. Agents Richardson and Barnard, who depended on interregional capital flows in their commission business, had better access to needed funds. As they had already made clear, however, they did not believe southerners would use the new technology, and thus they hesitated in investing in it.[38]

Southern agents' lack of technical expertise may have been a final factor behind their reluctance to organize exchanges. Only Burgwyn, who was an engineer by profession and claimed to have studied electricity, had a technical background. Despite his training, however, he did not understand the nature of the technological revolution telephony was undergoing. Though he experimented, or claimed to, with microphonic transmission, he believed that problems with telephone volume could best be solved with a headset.[39] He thus did not foresee the invention of the Blake transmitter, which greatly improved transmission quality. Or, if he did anticipate this device, he preferred the simpler and cheaper expedient of headphones. The driving force behind telephone growth, however, involved the shift to more complex and expensive technology, such as telephone exchanges. These innovations ran counter to the Virginia agent's preference for simple, cheap equipment.

There is little doubt that the problems of telephone promotion in the South were much greater than in even the most remote parts of the North. Poor and overwhelmingly rural, the South was still recovering from the ravages of war and completing the adjustment from a slave to a free labor economy. Its per capita income was only about half that of the rest of the nation between 1880 and 1900. Even though the region grew, it was unable to catch up with the more advanced parts of the country; by the end of the 1920s southern per capita income still stood at only 55 percent of the national average.[40] The South also remained the least urban and industrial section of the nation. Most of its widely dispersed urban population lived in what would have barely been termed towns in the North, while its unstable agricultural economy was characterized by land tenancy, rural isolation, and a shortage of cash.

Nonetheless, as bad as conditions were in the South, they cannot fully account for the region's poor response to the telephone. Conditions in many parts of New England were also unpromising at first, but aggressive Bell field agents brought the telephone to them. In large cities, they fit the telephone to patterns of demand for existing communications systems. In smaller cities near large metropolises, they discovered patrons among shippers, wholesale merchants, and especially regional manufacturers who needed lines directly to larger commercial centers. If the South had fewer markets of these types, it did have small towns, like the New England hinterland. In these places southern entrepreneurs could have found willing customers among bankers, coal dealers, general merchants, grocers, pharmacists, and other local businessmen who needed local telephone service in their work. Clearly telephone demand was lower overall in the South, but it was not wholly absent. By moving into the new towns and cities of the New South, southern promoters could have found the customers they sought.[41] As we will see, demand for the telephone existed in the South, but it took an outside entrepreneur to cultivate it.

Nor was the shortage of capital a fatal problem for the southern agents. In the early stages of telephone promotion, only credit and working capital were needed, not huge amounts of fixed investment.[42] Private line customers set up their own poles and wires, and before the exchange, there was little central office equipment to consume capital. Well-to-do cotton factors in port and terminal cities or rising upcountry merchants could have met these needs easily. The exchange presented more difficulties, exceeding the personal resources of most agents. Still, men like Richardson and Barnard, if they did not have the money themselves, were capable of raising it through their northern contacts in their mercantile activities. But neither they nor anyone else in the South came forth with the resources the industry needed.

If these conditions alone do not account for the initial retardation of the South's telephone industry, then at least some of the problem may have been due to a failure of entrepreneurship. Entrepreneurship here means the ability to innovate, to successfully introduce a new product, process, or technique in the face of adverse market conditions. Conditions certainly were adverse in the South, as they were elsewhere in the nation in the early years of the telephone industry; in contrast to other areas, however, entrepreneurs in the South either ignored the telephone altogether, or made ineffectual efforts to deal with the problems of introducing the new technology to their region.

In his classic study of entrepreneurship in economic growth, Joseph Schumpeter argued that entrepreneurs are motivated by the search for unusual or above normal profits.[43] Where such profits exist, or are thought to exist, entrepreneurs will act. As Schumpeter recognized, the individual's perceptions of profit, risk, and potential earnings must play a key role in such motivations, for at the beginning stages of an industry there is little concrete information available on actual returns. While in theory entrepreneurs can compare the expected return in a given undertaking with those of all other existing and potential endeavors and act accordingly, in fact they have only limited knowledge on which to estimate future outcomes and compare options. Calculations of rates of return would have had little meaning for the early telephone agents, for example. Their "investment" in the telephone industry consisted mainly of their time and energy. However well or poorly they may have done financially during their first few years, they had no way of knowing what the future might bring once the telephone was more developed and well established. This is not to say that profits and rates of return contribute nothing to entrepreneurial motivations, only that individual entrepreneurs respond to economic conditions through a filter made of their own perceptions, values, and beliefs. An entrepreneur contemplating entry into an untested business venture will not act on broad comparisons of returns in many fields, but on more limited considerations of what he believes he might earn in a smaller number of fields of endeavor congenial to his talents, resources, and preferences. Nor will calculations of short-term profits alone determine entrepreneurial behavior. Rather, those who have sufficient faith in a new technology to enter the business on the ground floor with the hope of rising as far as it will take them will become the first agents.

Because entrepreneurship means the successful introduction of an innovation, it depends not only on the subjective motivations of individuals, but on their particular talents and business skills as well. Such skills may consist of many things: salesmanship, technical knowledge, connections to sources of capital. Generally, at the early stages of an industry, perceptions and creativity are important. The ability to conceive of uses for a new technology, to find means around adverse conditions, and even to persevere in the face of competitive threats and initial failures may all be important, since early on there is not a ready-made market for a new product.

Both the entrepreneurial motivations and entrepreneurial skills of southerners seemed to have contributed to the initial problems of introducing the telephone to the South. Adverse perceptions of the market potential of the telephone, for example, may help to explain why so few southerners bothered to enter the new industry. Without an examination of the minds of a wide cross section of the southern business community this assertion cannot be proved.[44] We can, however, examine the behavior of southern telephone agents for evidence of how motivations and perceptions may have affected southern entrepreneurship.

Early southern agents approached the telephone with great caution, which suggests that they had low expectations of the market for telephones. Indeed, Cary apparently misperceived the market altogether, ignoring important centers of demand at a time when telephone use was becoming concentrated in central places.[45] Richardson and Barnard, fearing demand for the new technology in the South was light, also moved slowly, unwilling to make the personal investment of time and money the telephone required. Even the invention of the exchange failed to change their minds. The agents blamed the capriciousness of southern consumers and their lack of capital. Yet they failed to exploit opportunities that did exist. In response to Gardiner Hubbard's early suggestion that they form a company to organize regional promotion, Richardson and Barnard responded, "you know there is a great difference between the manner of doing things here and in the North and where it would be easy to organize a company in the North, it would be hard here."[46] Convinced that the South was different, and unfavorable for the telephone, the agents were unwilling to commit more resources to the business.

The underdevelopment of the South clearly contributed to and reinforced these entrepreneurial perceptions. The region had few insurance, express, or telegraph agents, men who could have moved laterally into the new industry. Those it did have, especially outside of large cities, were typically involved in other activities. One need only think of the upcountry merchant who acted as credit merchandiser, cotton ginner, express agent, and postmaster for his hamlet. Such men were secure in their positions and had little incentive to spread themselves thinner still by tackling a suspect new venture such as the telephone. Without an existing base of industry, the South also had fewer men of technical skill and training to assist with the new technology of the exchange. And capital, though available, was sometimes hard to come by, reinforcing the reluctance of agents to undertake the larger-scale investments telephone exchanges required. But it was the combination of adverse conditions and entrepreneurial perceptions that finally blocked the growth of telephony in the region.

The impact of these twin forces is perhaps most clearly illustrated in the case of C. P. E. Burgwyn. Finding southern reception to the new technology lukewarm at best and unsure of his future prospects, Burgwyn refused to invest the necessary resources in the project to make it a success. Lacking sufficient capital himself to support the telephone, short on even modest calls for funds, he never succeeded in enticing new investors into the industry, as did northern agents. Without working capital, his early disappointments sent him on a downward spiral, confirming his worst fears. The exchange exacerbated these problems. Requiring a greater commitment of capital to a more complex technology, it was even less suited to Burgwyn's underfunded, seat-of-the-pants type of operation. Objective conditions reinforced subjective impressions, making him even more reluctant to commit himself to the telephone and leading to his failure.

A similar mixture of external conditions and individual perceptions contributed to the inability of other southern telephone agents to find entrepreneurial solutions to the problems of introducing the telephone to the South. Effective telephone promotion took time, money, and, after the invention of the exchange, a certain amount of technical skill. It also required a sanguine view of the future. Southern agents possessed neither these resources nor this outlook. They operated as traditional businessmen, attempting to combine telephone promotion with their other locally based interests. But aggressive salesmanship was crucial in the early years of the business, for customers had to be convinced of the telephone's worth. The experiences as well as the basic business precepts of southern entrepreneurs dictated caution, but this was not what the telephone required either. New England agents who "pushed" the device gambled on future success. They gave free telephone demonstrations, allowed customers to use telephones before contracts were signed, and waited months for payment. Richardson and Barnard, by contrast, tried to develop the new business with a minimum of resources. To avoid paying Bell its advance on rented instruments, they rarely ordered telephones until they had secured contracts.[47] This policy kept inventory investment down, but also slowed telephone diffusion.

On the basis of the slim record left by southern agents it is exceedingly difficult to know for sure if their activities were the product of market conditions beyond their control or if the limited market for telephones stemmed from their lack of effective action. The agents themselves consistently pointed to the inherent problems of introducing the telephone to the South; a closer examination of conditions in the South and the behavior of southern agents suggests that poor entrepreneurship may also have been a significant factor. As has been argued here, entrepreneurship means innovation in the face of adverse conditions. Lacking the necessary entrepreneurial skills and motivations for this sort of undertaking, southern telephone agents could truthfully respond that there was nothing more they could do for the telephone. Their assumptions were quickly shattered, however, by the arrival of an outsider with a far different understanding of the potential of the telephone in the South.

Chapter 2

An Injection of Entrepreneurship

Dissatisfied with the response of southern agents to the exchange, and the generally poor state of the industry in the region, Bell managers sent a new man to the South. In late 1878, James Ormes arrived in Virginia to replace C. P. E. Burgwyn and to serve as general southern agent for telephone exchanges.[1] But his most important task was to pioneer a new approach to telephone development in the South. He was to assume the risks that southern agents had avoided. Taking up this challenge, Ormes achieved a remarkable record of success that revolutionized the market for telephones in the South.

Working with Ormes in his mission was James Tracy, a friend of the new agent and relative of Gardiner Hubbard. Arriving in Virginia, Tracy reported to Hubbard that chances were good for starting an exchange in Richmond. Though he found customers somewhat reluctant, he was convinced that with the right sort of intense effort he would succeed.[2] Like other agents, however, he needed capital. Reminding Bell managers that he had no funds of his own, he wrote, "I should like to be informed how to arrange the finances. Of course, the subscriptions will remedy that in a season, but while starting where am I to look for assistance."[3] As the parent firm still had no money to extend to its agents, the only answer was local sources. James Ormes made his way to Richmond less than a month later, sent by newly appointed Bell general manager Theodore Vail to help Tracy in this task.

Ormes was not a man whose character or background suggests he was the ideal choice for his arduous mission. In physical appearance he cut an unimpressive figure. A chronic sufferer of rheumatism and bronchitis from a young age, his ailments were magnified by the southern climate, which brought on debilitating attacks of hay fever.[4] While the southern climate bothered him physically, its culture and recent history must have pained him spiritually. A Civil War veteran, ardent supporter of the Union, and admirer of Lincoln, he had gone to Ford's Theater while stationed in Washington to catch a glimpse of the president, only to see his hero assassinated.

Ormes's position demanded he work within southern culture, and his record of success suggests he was effective in his business role regardless of his personal feelings or beliefs. If his war experiences or Lincoln's death gave Ormes any distaste for the South or animosity toward southerners, he did not show it. Indeed, as he wrote to his father-in-law after arriving in Richmond, "The house where I live is full of ex-Reb. generals, but they are really very affable and I never had a more enjoyable season in my life than my past few days here."[5] He quickly impressed southern businessmen as well. Soon after his arrival in Richmond, one wrote, "he is comparatively a stranger here, but seems to be a smart, active, clever, young man."[6]

Ormes was in fact stranger neither to the South nor to the position of agent for northern interests in a remote territory. Though born in Brimfield, Massachusetts, his life in the Union Army had taken him to Washington, D.C., and later to Falls Church, Virginia, providing him with some familiarity with the territory he would serve as Bell agent.[7] Since the war, he had been a promoter of far-flung business ventures, serving as agent for a variety of New England firms in the West. In this sort of work, Ormes drew on his inner resources, particularly his strong religious convictions, which helped him to overcome the trials of his labors. Away from family and friends, his profound faith provided him with the force of will he needed to undertake difficult and at times physically painful tasks. As revealed in his correspondence, feelings of love for his family, wife, and God sustained him as he moved about the country in the lonely task of promotion.[8] He would draw on this faith during his time in the South as well.

Very quickly Ormes impressed Theodore Vail at the Bell Company. As the general manager made clear to Ormes from the start, he would be given every opportunity to take command of the southern agency. "It is our desire to give you the territory entirely," Vail wrote, "and if your efforts are successful, we shall cancel the existing contracts with other parties."[9] The company's executive committee concurred with Vail's assessment, agreeing that if Ormes could develop the southern territory and start exchanges, he was entitled to an exclusive franchise.[10]

With this encouragement, Ormes got right to work in Virginia. Following the same basic strategy that Burgwyn had pursued, he succeeded where his predecessor had failed. The new agent interested prominent public officials and businessmen in telephones, paving the way for further rentals. By March 1879, he was preparing a contract for telephonic service in the Richmond Tobacco Exchange.[11] So vigorous were his early efforts, in fact, that Vail reminded him not to spread himself too thin or lease telephones without firm contracts.[12] Ormes persisted, however. Only days after his arrival in Virginia he had ordered a switchboard from the Baltimore electrical manufacturers Davis and Watts, though he had no exchange to use it in.[13]

The northern promoter successfully blended his two tasks—renting telephones and preparing the South for the introduction of the exchange. His "vigor and push" produced rapid telephone growth in the region, surpassing the parent company's expectations. As Gardiner Hubbard dryly observed, "the value of the southern territory proved to be very much more important than any of us supposed."[14] With somewhat more exuberance, Ormes confidently assured Theodore Vail that he would have over 300 subscribers for an exchange in Richmond.[15] As general agent, Ormes also furnished Bell executives with reports on southern business conditions and monitored the activities of his fellow agents.[16] These administrative functions would grow more important as the exchange business expanded.

Initially, however, Vail and the other Bell managers were most impressed with Ormes's aggressive activity in the face of a growing Western Union challenge.[17] No mere pretender, the telegraph firm was the largest corporation in the nation at that time.

It possessed vast resources, a well-known name, and its own patented telephone. The giant's influence with railroads had frustrated Burgwyn, and its presence in telephony continued to grow throughout 1878. Still, Western Union was relatively weak in the South. As Bell managers recognized, if the interloper could not be defeated in this market, there was little hope of beating it in its strongholds in the Northeast and Midwest. For this reason, Vail watched closely Ormes's duel with his rival.

Like other agents, Ormes was at first reluctant to take on Western Union and appealed to Bell for help. Vail's advice was to meet competition with competition, advertising circular with advertising circular.[18] The parent firm could not afford to take legal action against Western Union encroachment; nor was Bell sure its telephone patents would stand up in a long, drawn-out court fight.[19] By June, Ormes grasped the seriousness of the situation and suggested to Vail that Bell help him undercut their adversary by employing the telegraph firm's own agents. Vail refused, stating that it would be impossible to do so without voiding Ormes's own contract.[20] Though the general manager added a vague reference to some negotiations with Western Union—a hint at the eventual 1880 settlement—he gave no firm promise of help and Ormes was left on his own.[21] Bell was depending on Ormes to take whatever action necessary for a victory in the South. Placing great faith in their southern operative's innovative talent, the parent firm encouraged him develop his own ideas.

Under a severe competitive threat, the northerner responded in an innovative fashion. He began by creating a formal organization for his promotional efforts. In early 1879, Ormes formed a partnership with Daniel I. Carson. A printer by trade, Carson had met Ormes years earlier while working for a Chicago publishing house. In 1878 Carson moved to New York, where Ormes contacted him about his southern telephone agency.[22] Drawn to the Bell agent by their common religious devotion, Carson accepted the position as Ormes's private secretary and business manager, investing $600 of his own money in the partnership and agreeing to take one third of the profits in lieu of a salary.[23]

Though the partnership provided some of the support Ormes needed, he still had to tap local resources to build exchanges. In the North, Bell promoters could rely on regional manufacturers for monetary support. Ormes had to draw funds from a different economic elite. As he wrote to Carson, he would have to begin by cultivating the support of "prominent men" for his efforts, then involve judges, lawyers, fire and police chiefs, merchants, politicians, and finally, general businessmen.[24] In this endeavor, Ormes proved successful, in contrast to his predecessor Burgwyn. Though he found Richmond businessmen "conservative," he carefully solicited their support for the city's telephone exchange by offering nine local leaders free stock for their service as directors of an exchange company.[25] Aware of the value of organization, he tried to form a syndicate with these men to mobilize $25,000. As the northern agent soon discovered, however, southern businessmen were far more forthcoming with encouraging words than capital.[26] Indeed, they responded to his pleas for aid with indignant assertions of their supposed "rights" in the Richmond exchange, though none of them had actually paid in any money.[27]

That Ormes refrained from abandoning all hope of reliance on local support is a mark of the man's determination and self-restraint. Eventually he was able to organize the syndicate he wanted. John Branch, a Richmond stockbroker and the president of the Merchant's Bank, John Chamberlyne, editor of the local newspaper, Philip Haxall of the Haxall and Crenshaw flour mill, and a collection of merchants, bank officers, physicians, and professors formed the Richmond Telephone Exchange Company in May 1879. Ormes's skills as a promoter served him well, for he was able to repeat his success with a similar organization in Norfolk.[28] Men such as Branch, financier O. T. Terry, and Lynchburgh bond dealer A. M. Davis had access to the sort of funds Ormes needed. By the spring of his first year in Virginia, the Bell agent had uncovered the few investors in his territory with the resources and foresight to support telephony.

Just as important as Ormes's connections within the South were his strong ties to the North. These connections provided him with capital as well, as he drew on Boston friends to build the Richmond and other exchanges.[29] Through correspondence to Boston, he also kept in close contact with Bell agents in New England, who provided him with useful technical information. From George Coy, inventor of the switchboard, he learned early on of advances in central station technology.[30] From Lewis W. Clarke, Bell agent in Rhode Island, he heard of a new signal bell in use in the North.[31] Business information also flowed along these channels, in both directions. Ormes gained from Bell agents across the country ideas for advertising the telephone; in turn he provided George Maynard in Washington, D.C. with advice on running exchanges.[32]

With access to these resources, Ormes moved beyond Virginia and proceeded to promote telephone exchanges throughout the South, much in the same fashion as did New England agents. With instructions from Thomas Watson, he began by setting up district telephone and fire alarm systems in southern towns. These systems, an intermediate step between private lines and full-fledged central office operations, relied on the same principles as the exchange—customers were hooked in sequence to a central point—but were simpler and cheaper. Telephonic fire alarm systems allowed citizens to call in an alarm from a number of points in town. The district system worked in a similar manner. Calls to a central office were relayed to other points via messengers and telegraph. District service was especially useful to merchants and other town businessmen, men who would find exchanges attractive as well.[33] By June 1879, district systems were in place in Petersburgh, Virginia and Wilmington, North Carolina, preparing the way for the exchange operations that would commence some months later.[34]

Although these successes were encouraging, Western Union still threatened to overwhelm Ormes's fledgling enterprise. In April Ormes had tried using price competition and attacks on Western Union's size and market power, portraying Bell as the underdog, a "home" company battling a foreign giant.[35] But neither tactic was effective. Though he reduced Bell rates to $32.00 per year in Lynchburgh, many citizens still opted for Western Union telephone service at $48.00 annually. Ormes dropped rates to $24.00, but was forced to turn his policy around when the new price attracted few customers.

As the summer of 1879 approached, Ormes shifted his competitive strategy. Recognizing that profits were to be found in opening exchanges in large cities, where the number of potential customers was high and the cost of constructing lines low, the Bell agent adopted an aggressive policy of urban expansion. In the most promising New South metropolises he fought Western Union by trying to convince customers that Bell service was higher in quality than that of its rival.

Between April and June, Ormes pursued this strategy vigorously, moving out from the confines of his Virginia territory. Taking a broad regional view, he sought markets throughout the South that could support exchanges. With his partner Carson, Ormes surveyed the entire seaboard South for customers.[36] This regional strategy proved to be Ormes's most important entrepreneurial move, for it eventually enabled him to surpass Western Union and capture all of the most important southern markets for Bell. It was also his most controversial action, however, for it brought him into direct conflict with existing Bell agents, Richardson and Barnard in particular.

Following the 1 April opening of the Richmond telephone exchange, Ormes executed exchange contracts for Norfolk, Lynchburgh, Petersburgh, Wilmington, North Carolina, and Augusta, Georgia. Most importantly, the entrepreneur moved into Atlanta, a city which Vail had presciently called one of the brightest prospects in the South.[37] The Bell general manager assisted Ormes with a special contract for this city which eased his cash requirements by providing a longer time to pay advances on telephones.[38] The agent responded by finding support for his project from leading members of Atlanta society. Congratulating Ormes on his achievements, Vail wrote, "I think the parties you have succeeded in interesting are the best possible parties you could have obtained."[39]

While Ormes and Vail were pleased with this progress, agents Richardson and Barnard were not. When the Atlanta exchange opened in June 1879, they realized that Ormes was threatening to undermine their territorial franchise, which they took almost as a vested right.[40] As general exchange agent for the South, Ormes clearly had a right to act in this fashion. But relations between the northerner and existing Bell agents had never been formally settled. The early contracts were vague and did not include provisions for exchanges. Informally, Vail allowed the older licensees to operate in their original territory, fearing that giving it away to others might bring legal reprisals. Under these conditions, however, disputes between Ormes and his fellow southern agents began to appear.

The Bell general manager used the vagueness of the parties' rights to mute conflict, while hoping that Ormes's aggressive tactics would spur Richardson and Barnard to embrace the new technology of the exchange or leave the telephone business. In June, he implored the Georgians to "work up" their territory and pursue a more "active" policy. He requested more than once that they consider devoting full time to the telephone.[41] Apparently, however, Richardson and Barnard were barely doing the minimum necessary to fulfill their Bell contract.

During June, Vail issued to Richardson and Barnard an ultimatum. Either they were to stop letting profitable opportunities in exchange service go by, or they should no longer consider themselves Bell agents.[42] Since private line rentals were declining as a percentage of total business, refusal to move aggressively into exchange promotion was, in Vail's mind, tantamount to resignation. In the future, private lines would become adjuncts of exchanges, linking central offices with outlying areas. The agent who controlled the profitable central office business would have the capital, technical knowledge, and business contacts needed to dominate private lines as well.

Vail's admonitions apparently had little lasting effect on Richardson and Barnard. He continued to chide them for their failure to start central office operations in Savannah, writing, "I am very sorry that the prospect for a district or exchange system in your city is not better and that you do not see enough in it to encourage you in going ahead."[43] Vail then raised the specter of competition, adding, "I will write to Mr. Ormes and he will probably see you at an early date and you then can decide what is best." Though Vail may have believed the Savannah merchants could still make a worthwhile contribution to the telephone business, it appears he was hedging mainly for legal purposes. That same day he wrote to Ormes, "now is your chance to go in and occupy the territory."[44] Richardson and Barnard did begin to push the telephone with somewhat more vigor, and, overcoming their initial skepticism, they eventually opened exchange service in Savannah.[45] But for Vail, Richardson and Barnard's efforts were too little, too late. The general manager had already made up his mind that they could not be trusted. Bell's strength lay in its still shaky control over telephone patents, and Vail feared that if instruments fell into competitors' hands, the patents would be violated. Independent-minded agents such as Richardson and Barnard did not inspire the general manager's confidence. His distrust probably grew even stronger as he began to perceive that future telephone growth depended on a system of central exchanges linked by intercity lines. Vail wanted to reserve for the Bell Company the right to make the vital interexchange connections.[46] To carry out these plans, he needed loyal agents who were willing to take a broader view of the industry, or who at least would follow instructions. Richardson and Barnard's narrow-minded resistance to new technology and overweening concern about their franchise rights boded ill for the future.

With Vail's distrust of the Georgia agents mounting, he came to depend even more heavily on James Ormes. As he recognized, the Virginia agent's control of exchange development would allow him to determine the future path of the southern telephone industry. Advising Ormes to keep his relations with Richardson and Barnard cordial, Vail made it clear that he still hoped to make the northerner the exclusive Bell agent for the South.[47] All that stood in the way were some tricky legal matters. To surmount these, he asked Richardson and Barnard to return their correspondence with Gardiner Hubbard, so that he might know more exactly the terms of their contract.[48] In fact, Vail was searching for a legal loophole to terminate their franchise. Ormes assisted the Bell manager in this effort by avoiding entanglements with other agents and tightening his hold on important southern markets.[49]

By 1880 Richardson and Barnard's connections with Bell came to an end. There were no dramatic confrontations; rather, they were squeezed from the telephone industry by Ormes's control of the predominant exchange business. Ormes and Vail provided from the outside the innovative effort that the South could not generate on its own. By expanding the role of the parent firm and enforcing tighter top-down control, Vail circumvented Richardson and Barnard's resistance to change. By successfully moving into the exchange business and developing the Virginia territory where others had failed, Ormes exploded their assumptions about the southern market.

The importance of Ormes's efforts can be clearly seen through a comparison of the southern and northern telephone industries. In the North, too, innovative efforts were needed to complete the transition from private line to exchange service. New England faced many of the same problems, conflicts, and potential restrictions on growth as the South. Residents of the smaller New England towns greeted the exchange with skepticism. As one Massachusetts man replied to agents Stearns and George's suggestion that he join a district company, "I do not want *any* public management of my private business."[50] Overcoming such objections fell to Bell telephone agents. In contrast to the South, however, New England had a large and active entrepreneurial class to draw on.[51] This group performed the same functions that Ormes performed virtually single-handedly.

New England's many field and subagents had already proven themselves invaluable in private line promotion, and they were no less important to the progress of exchanges. As a result, their status began to rise at the expense of general agents after 1879. When the telephone was new, general agents such as Stearns and George and William Hayward had been ideal middle-level functionaries in the Bell agency structure. At that time they held sufficient capital and credit to serve private-line customers. But they did not have the funds necessary to meet the requirements of a territory full of exchanges.[52] To fulfill these needs agents had to find new sources of capital. This work could best be done by their field and subagents, who had made contacts with local businessmen in promoting private lines. They found investors, raised capital, and sold the new system to consumers.

As in the South, such changes bred resistance from older agents. Some of the licensees appointed by Gardiner Hubbard clung to a conception of the telephone formed in the days before exchanges. Hubbard had stressed the decentralized nature of telephone promotion, and gave agents autonomy to organize their territories as they saw fit. The Bell Company, as holders of the telephone patents, expected to reap the rewards of their efforts but to participate very little in the actual work of promotion. Under this regime, a certain set of relations had arisen between licensees, customers, and the parent firm. Agents acted as semi-independent intermediaries between Bell and their customers. Taking a proprietary interest in their territories, they resisted encroachments from others and fought to maintain their autonomy.[53]

Contentment and complacency were especially noticeable among the general agents of New England, who saw themselves as independent businessmen in charge of a group of employees—their subagents. Some general agents were not willing to abandon this view, even in light of the new demands induced by the exchange. William Hayward, for example, resisted change so vehemently that he was eventually forced out of the industry. Upon entering the telephone business, he had negotiated strenuously for as large a territory as possible, adding the telephone to his other business interests.[54] In the early days of telephony, he had operated from a position of strength, and was able to work out of his Norwalk, Connecticut and New York City offices, overseeing the efforts of the subagents of his territory.[55] Hayward tried to protect his franchise by enforcing the provisions of his contract that gave him exclusive rights in his territory. Such efforts stifled innovation and lowered rates of growth, but they also assured Hayward that he would garner profits from a wide field once the telephone business took off.

Legal barriers proved no match for the demands of the market, however, as Hayward found his agency slipping into the hands of his subordinates. Between 1878 and 1879, George Coy and Thomas Doolittle, with the assistance of Boston financier Charles Cheever, took over the key entrepreneurial functions in Hayward's territory.[56] They made the direct contacts with investors and subscribers necessary to build city exchanges, and organized a company to carry out their plans. While Hayward protested his loss of status, he found his strength undercut by the changing structure of the industry. He received compensation in the form of stock in the New Haven exchange, but his agency was disbanded and his franchise transferred to the newly formed New England Telephone Company.[57]

Though his role in the telephone industry was unquestionably reduced, Hayward actually profited from the change. In the long run, the payment he received far exceeded the value of an obsolete territorial franchise. Yet to Hayward, this shift in status seemed at first a loss. Like most early agents, he had been attracted to the telephone by the future profits he expected to extract from a valuable property right—his Bell franchise. He was unwilling to abandon this vision, even in the face of changing conditions. While he saw that the exchange brightened his prospects for profit, he failed to perceive that it also negated his former role. In this way of thinking he was little different than Richardson and Barnard, who believed that their Bell franchise was a vested right that they could use as fit their business habits. Only over time did such men come to see that a moderate-sized town using the telephone exchange could generate substantial profits. It took other entrepreneurs, operating from below, to force a revision in their outlook.

This creative impetus came from within in the North, while it came from without in the South. Even some early Bell general agents in New England rethought their business strategy, stopped fighting change, and sought out new opportunities. Cozzens and Bull, agents in Rhode Island and Connecticut, understood what was happening in telephony and began building exchanges in their territory. They also tried to preserve their autonomy by maintaining an intermediate role between their customers and Bell. Purchasing their own equipment to start central office operations in Norwich, they tried to negotiate favorable rates with the parent firm.[58] Though Bell managers favored exchange promotion, they did not want to cede proprietary control over the telephone business to agents. Theodore Vail had envisaged an industry composed of a number of exchange companies organized by Bell representatives and local capitalists, but firmly under the control of the parent firm. He thus rejected Cozzens and Bull's initiative and forced them to maintain standard Bell rates.[59] For similar reasons Vail also quashed their attempt to create a regional long-distance telephone company. The agents tried to reserve for themselves the right to make interexchange connections. Had they done so, they would have been able to oversee a system of local exchanges by connecting them through long lines, as Bell itself would do later. But Vail had already foreseen this possibility and wanted to preserve the long line business for the parent company. He thus refused their request to put up a line between Norwich and New Bedford, and prevented them from establishing an independent telephone company remote from Bell's control.[60] As this episode demonstrates, however, the general manager had to restrict rather than encourage aggressive responses from New England entrepreneurs.

Through efforts such as these, New England agents on their own reorganized the region's telephone industry. New England had a wealth of technically trained, aggressive promoters who quickly perceived the advantages of new technology. Some were men of striking talents. Charles Cheever, for example, was an original thinker who early on saw the need for large organizations capable of mobilizing huge sums of capital for development.[61] Thomas Doolittle was a technical virtuoso who applied his acute mind to the relationship between technology and organization. He was one of the first to use the word "system" to describe the evolving network of local exchanges and long-distance lines.[62] Often without funds themselves, these creative entrepreneurs found support for their ventures among local businessmen and innovated as they went along. In doing so they revolutionized the nature of telephony in New England.

Comparisons with the North make clear some of the reasons change and innovation came more slowly to the South. From the very beginning the pool of telephone agents, totalling some sixty-five individuals between 1877 and 1880, was greater in New England than in the South, which had only three members plus Ormes during the same years. In addition, few capitalists in the South were willing to risk their funds in telephone promotion, in contrast to the North where agents found support for their work among local businessmen, especially manufacturers.[63] With the exception of Atlanta, and to a lesser extent Richmond, southern businessmen showed no more than a passing interest in the telephone, a problem that continually troubled Ormes.

Ormes's success in the South also demonstrates that southern agents were simply wrong in their belief that the market would not support a greater promotional effort. Southern telephone entrepreneurs may have been put off by the lack of major cities and manufacturing towns in the region such as existed in the North, where almost any agent could see that demand for the new technology was strong and would grow in the future. Still, Ormes came to this same conclusion in the South. Overall demand for telephones may have been smaller in the South than in the North, but, concentrated in a few growing cities near the South's major manufacturing and marketing centers, it did exist and should have received at least some assistance from indigenous southern businessmen.

Ormes's perceptions of risk and the market differentiated him from his fellow agents in the South. Rapidly rejecting outmoded methods, quickly embracing new technology, the northern promoter injected needed entrepreneurial vitality into the southern telephone industry. Where southern agents were cautious, Ormes was bold. Where they took a restricted view of the market, Ormes devised plans to open up southern telephone demand. By contrast, southern businessmen such as Richardson and Barnard continued to believe that southerners were too conservative and the southern market too thin to generate a strong demand for telephones and telephone exchanges. They therefore failed to make what on the whole seems a reasonable trade—their rapidly declining private line franchise for an exchange contract in a major city. With the future unknown, this move was unclear and risky; and, as William Hayward demonstrated, northern entrepreneurs sometimes thought in the same way. But in the South, such reasoning seems to have had a pervasive grip on the business class. These perceptional barriers had inhibited southern investors from supporting the telephone at its early stages, when it needed their help the most. As the industry progressed, this same aversion to risk and misperception of the market continued to dissuade them from committing resources to the enterprise.[64]

Uninhibited by the considerations that affected southern entrepreneurs, Ormes undertook the sort of intensive promotional effort that the telephone needed. His approach suggests he had perceived that the diffusion of new technologies generally follows an S-curve: growth begins slowly until it reaches an inflection point where it takes off rapidly. Competition among places for trade and commerce can act as a stimulant to rapid diffusion. On this basis, Ormes adopted a regional strategy that fit this pattern of diffusion. Mobilizing capital from relatively prosperous places such as Richmond, he used it to move beyond the confines of his original territory and seek the most promising markets for the telephone throughout the South. Success in cities returned profits that fueled further growth, generating a demand for telephones that spread the device to the hinterland. Following this strategy, Ormes did on his own what numerous individual agents did in New England and elsewhere.[65]

Important as Ormes's achievements were, they were not sufficient to ensure the continuing vitality of the southern telephone industry. Telephony was still undergoing rapid change, as the growth of exchanges and the beginnings of long-distance service called for more capital and greater organization over a larger market. In New England, individual agents, joining forces and drawing on regional supplies of capital, were beginning to carry out this next stage of telephone development. The South still had only Ormes and his small entourage to rely on; despite his accomplishments, he had not completely broken the perceptional constraints on southern entrepreneurship. Southern businessmen's outlook and behavior continued to bedevil him and frustrate his plans for building the southern telephone industry. His old nemesis, Western Union, also loomed large in the background. Though temporarily put on the defensive, it had not been defeated. Ormes needed still greater resources and tighter organization of his southern agency if he was to combat the relentless Western Union machine. In this effort, the northerner would continue to find little assistance and much resistance from southerners. He was forced to turn elsewhere for the help he needed.

CHAPTER 3

An Organizational Solution

BY MID 1879, Ormes had achieved the results he promised Theodore Vail when he took charge of the southern agency: he had found a market for Bell telephones and he had successfully introduced the telephone exchange to the South. Still, with Western Union continuing to threaten, Ormes almost lost the support of Daniel Carson. "*We can* and shall clean them [Western Union] all out," he wrote to bolster Carson's wavering confidence.[1] Carson stayed, ending one danger, but others persisted. Capital was still short and resistance to change by southerners threatened Ormes's unsteady enterprise.

The shortage of capital was perhaps the most serious problem, for it prevented Ormes from competing effectively with his wealthier adversary. Since the advent of the exchange, capital costs in telephony had increased dramatically. Basic switchboards ran $150.00 at the outset. Annunciators cost $5.00 and bells between $5.00 and $10.00 each, significant sums in an exchange of 50 or more circuits.[2] Northern agent George Coy informed Daniel Carson that the fixed cost of an exchange would total more than $5,000, and annual operating expenses could be expected to reach $3,500.[3] Expenses such as these well exceeded Ormes's personal means and had led him to form exchange companies in Richmond and Atlanta.

Despite success in these cities, southern investors continued to spurn the telephone industry.[4] Trying to convince O. T. Terry of Virginia to support his undertakings, Ormes advised his ally Carson to push matters as quickly as possible and keep this important investor interested: "I'll be able to pass in some cash, till then, I tell you, boy, you must root hog or die."[5] Yet such efforts were not always successful. Other local investors who had taken an interest in the Richmond exchange were not willing to supply further capital for ventures that went beyond their immediate locality. Nor could Ormes find dependable supplies of cash outside of the South. In late May, John Munson, a Boston stock dealer on whom Ormes had hoped to depend, wrote that he could not supply needed funds.[6] By June, the situation was growing desperate, as Ormes was pressured by his subordinates for the cash they needed to press on with their work.[7]

New troubles with southern entrepreneurs added to Ormes's growing list of worries. Local interests who had initially supported the Atlanta exchange reversed course and conspired to take control of this valuable property from Ormes and Bell, touching off a bitter legal struggle. This episode must have been especially disheartening to the northerner, for Atlanta had seemed the great triumph of his early career. The growing New South metropolis, famous for its civic boosterism, had at first responded warmly to Ormes's promotional efforts and contacted him in March about opening an exchange in their city.[8] Ormes, along with Joseph E. Brown, a lawyer and ex-governor of Georgia, and his son Julius of the important Western and Atlantic Railroad, each took one-third interest in a new firm. Joseph Brown was just the sort of man Ormes thought he had been looking for to aid him. Reputed to be one of the wealthiest men in the state, Brown had numerous regional business interests and strong political connections.[9] With the Browns' help, Ormes was able to open an exchange in the city in 1879. Reflecting with satisfaction on the speed of progress in Georgia, Ormes invidiously contrasted the progressive New South city with "conservative old Richmond, [where] the fixed capital stays fixed."[10] Theodore Vail seconded Ormes's confidence with an exceptionally favorable Bell contract for the city to help assure success.[11]

Both Vail and Ormes had cause to regret their satisfaction,

however, as trouble soon broke out. The Atlanta Exchange Company was only tenuously connected to the rest of Ormes's agency. To reward local participation, Vail had given the Atlanta company a contract that made it a small but important focal point of regional telephone development. In sixty miles in all directions, the organization was allowed to rent telephones and provide service, enabling it to embrace important regional marketing centers such as Rome, Marietta, and Griffen. With such a contract, the Atlanta company operated in semi-independent fashion from the rest of Ormes's domain. The company, not Ormes, held the Bell franchise for this territory.[12] Such was the nature of the early, decentralized telephone business. Conflict came when Ormes tried to link the Atlanta company closer to the rest of his agency.

Trouble first arose over legal and financial matters.[13] The Georgians claimed that they were entitled to a greater say in managing the property, as they had put more capital into it. Ormes disagreed, noting that the nominal figures of stock subscription aside, he had invested ten times as much as they had in the process of cultivating a clientele for telephone service. This response did not quell the controversy. Ormes's partners also maintained that under the terms of their franchise, they had a right to develop by themselves the large territory surrounding Atlanta. Ormes, they claimed, was infringing on their rights under this provision by continuing to pursue his strategy of regional telephone promotion.[14]

While it is not clear when these problems arose—Ormes reported in July that he was anxious and confused over Atlanta—the real battle commenced in the fall of 1879.[15] Between September and December, the conflict grew until mutual antagonism dissolved the partnership. Ormes characterized his southern partners as corrupt politicians more interested in bilking a "yankee" enterprise than in renting telephones. Part of this charge was true. The Browns were a powerful political family in the state. Joseph Brown had been the Confederate governor of Georgia, in which office he was renowned mainly for his refusal to follow dictates from Richmond. He was also a notorious political and economic opportunist. After the war he quickly switched allegiances from the Democratic to Republican party, until the defeat of the Republicans returned him to the Democratic fold.[16] A few years before his plunge into telephony, Brown had been involved in another controversy over a paper mill in Marietta, where he fought a similar battle with his partners for control of that enterprise.[17]

In the great Gilded Age tradition, the Georgians saw their opportunities and took them, or at least tried to. But the underlying cause of the Brown-Ormes conflict went deeper than a struggle over property. The incident was the result of the clash of the two incompatible notions of entrepreneurship held by the two parties. The Browns affected a cautious style of telephone promotion, one similar to that of agents Richardson and Barnard. Like other southern investors, they feared spending money aggressively on telephone development. Placing security above all else, they moved slowly in building up the Atlanta business, frustrating Ormes and the Bell Company.[18]

Though uninspired, the Browns' strategy did have a certain logic that could not be ignored. Familiar with business conditions in their area, they argued that few Georgia towns had shown an interest in telephones or telephone exchanges. For even a prosperous city such as Atlanta, moreover, adoption of telephones was a financial strain. The city fathers of Atlanta were cautious men themselves. Their municipality was already saddled with both a heavy bonded debt and $400,000 in short-term notes. They could not commit Atlanta to renting telephones for municipal use until the city was out from under this burden, as charter restrictions prohibited the city from going any further in the red.[19] Under such conditions, caution and conservatism seemed the reasonable course.

The Browns' reasoning, however, only fueled Ormes's anger. To him, it expressed their restricted view of the telephone market, a mindset he had already encountered many times in the South. The Bell agent saw Atlanta as the focal point of his promotional efforts in the lower South.[20] His strategy called for investing, at great initial cost, in multiple centers of telephone demand in the reviving southern economy. From vital towns such as Atlanta, use of the telephone would spread like ripples on a lake. Rome, Selma, and Athens would be next, if Atlanta was successful. In a sense, Ormes had conceived of a telephone network, though one linked by commercial, not telephone lines.[21] Trade and commerce between cities, he believed, would help diffuse telephones and exchanges throughout the South. Success in one place would breed imitation in others. Competition for business between cities would force all to adopt the telephone or suffer. From urban centers demand would also emanate to the hinterland, pushing the telephone deep into the southern economy.

Ormes's vision, ahead of its time for the South, seemed to the Georgians the dream of a yankee speculator unfamiliar with southern business conditions. They had their own ideas and tried to take control of Atlanta so that they could act on them.[22] The Browns, for example, took advantage of the decentralized nature of the early telephone business and their liberal Bell contract to earn side profits by renting instruments beyond the confines of Atlanta.[23] These efforts infuriated Ormes, who nonetheless continued to complain that his Atlanta partners were slow in promoting the telephone. But the outside rentals were not what Ormes had in mind. The Georgians tried to expand rentals piecemeal, casting about for parties interested in telephones, gradually extending their influence, moving only when the move would produce immediate profits. Ormes's more radical plans called for systematic investment in key centers of growth to generate a regional demand for telephones. This strategy thus cut across local boundaries and embraced the southern market as a whole, a policy diametrically opposed to the Browns' locally based efforts.

Knowing that their strength lay in Atlanta—where their political influence had secured a valuable right-of-way franchise—the Browns resisted Ormes's bold translocal strategy. It had, in their view, several drawbacks. It required a large speculative investment; it demanded allegiance to a faraway corporation; and it lessened their control of the Atlanta property by making it part of a larger enterprise.[24] While the southerners would not have minded extending their sphere of influence, they wanted to do so in ways which accorded with their existing lines of wealth and power. This involved drawing on their established contacts, which they tried to do by renting telephones in and around Atlanta. Like agents Richardson and Barnard, the Browns saw their investment in the telephone as one part of a portfolio of interests, a portfolio ground firmly in the Atlanta economy.

These basic differences of strategy turned the once hopeful Atlanta enterprise into a naked struggle for power. While originally it had seemed necessary to build local support for the telephone, localism had become, from Ormes and Vail's point of view, a hindrance to systematic telephone development. In response to the Browns' legal machinations, Ormes turned the tables and tried to railroad them out of the business. With Vail's support he surreptitiously invited Western Union competition in an effort to undermine the Browns' hold on Atlanta.[25] In the end the Georgians were defeated and this conflict marked the end of Ormes's efforts to rely on southerners for help.[26]

The Atlanta episode brings the differences between Ormes's style of entrepreneurship and that of southern businessmen into sharper focus. The contrast between Ormes's bold plans for regional telephone development and the more limited efforts undertaken by indigenous southern agents and investors suggests that localism may have been at the root of the South's poor entrepreneurial response to the telephone. The characteristics of the Southern economy, other writers have noted, tended to encourage investment in locally based projects.[27] Weak capital markets, high information costs, and high risk made the southern business class understandably cautious about placing their money in large-scale enterprises or ventures that they could not personally supervise.[28] To shield themselves from risk, they assembled portfolios of locally based assets, mainly land, to diversify their capital and cushion the potentially fatal blow of a sudden downturn in a single investment. The result was a strong link between the southern investor's property interests and the local economy.[29]

Though a perfectly rational response to conditions in the region, this localistic investment strategy did not provide southerners with the motivation to promote an enterprise like the telephone. The scale and scope of the telephone demanded a translocal strategy such as the one Ormes had adopted in order to diffuse the new technology throughout the scattered urban markets of the South. Yet investors in Atlanta and Richmond pointedly refused to embrace such a policy. Over time, the growth of demand and spread of the technology would reduce the risks of the telephone business and enable small-scale entrepreneurs to enter the industry; but as with many industries in the early stages, effective telephone promotion demanded a commitment of resources to a regional organization. In the South, this need was particularly great because the drastic shortage of telephone entrepreneurs made more decentralized approaches impossible.

Organizing a regional promotional campaign required a new sort of business mentality. The locally restricted, part-time undertakings of southern agents had to give way to the full-time, professional effort of a man like James Ormes. Unlike indigenous agents, Ormes calculated his profits strictly on the returns of rapid telephone growth, not as part of his portfolio of interests in the local economy. Almost as soon as he arrived, he made it clear that he would act in a greater capacity than just Bell agent for Virginia.[30] With his partners Daniel Carson and James Tracy, he tried to use local funds to seek out profitable opportunities throughout the region. This commitment violated what southern investors considered to be the principles of wise investment—keeping one's capital close at home and under one's direct supervision.

Convinced that southerners did not support and could not comprehend his entrepreneurial strategy, Ormes began to look outside of the South for the resources he needed. In the spring of 1879, even before the Atlanta partnership disintegrated, the Bell agent tried to alleviate his chronic capital shortage by establishing close contact with the Baltimore electrical manufacturer Davis and Watts. This strategically located firm served successfully as Bell agent in Maryland. It was also licensed by the parent company to supply other agents with telephone equipment such as bells, batteries, and wire.[31] Ormes apparently envisaged a mutually beneficial relationship between the manufacturer and his southern enterprise. Davis and Watts would provide telephones and telephone equipment, technical advice, and working capital to the South. In exchange, Ormes would guarantee the company a steady market for its output.

Ormes recognized the value of such an alliance in the rapidly changing telephone industry. With southern investors proving unreliable, and costs expected to rise exponentially with the number of subscribers, the agent needed a new source of funds. Unable to depend on the Bell Company for assistance, Ormes's most reasonable alternative was an outside manufacturing firm such as Davis and Watts.[32] Close ties to the manufacturer offered other benefits as well. Boston telephone supplier Charles Williams was pressed beyond capacity to meet Bell agents' needs. As a result, shipments of telephones to the South were frequently delayed, a costly situation with Western Union pressing hard into the market.[33] Williams was also forced to carry agents on his books for several months, until they obtained rentals from customers.[34] This situation strained his own resources and threatened to undermine the entire financial structure of the industry. By purchasing his telephones and equipment from a nearby supplier, Ormes could have helped to alleviate this financial crisis while placating his customers, who were "kicking and raising old nick" over equipment delays.[35]

In a similar fashion, Davis and Watts had the potential to provide the southern telephone industry with needed technical assistance. A shortage of trained engineers still plagued Ormes, forcing him to search desperately for men of technical skill, even if it meant hiring Western Union operators.[36] In the days before standardized design or large-scale research and development expenditures by Bell, agents had to make many crucial improvements and repairs of telephones and equipment themselves. Ormes, like other agents, had received instructions from the Bell Company so that he could do so, and the parent firm continued to send out technical information from Boston to encourage a cross-fertilization of ideas and a rapid diffusion of new technology. But for regional promoters located in remote and underdeveloped territories like Ormes, these efforts were insufficient. For those things Ormes could not fix himself he had to rely on engineers in faraway Boston. The shortage of skilled technicians in the South gave him little other choice. His only options were to return defective instruments, a time-consuming process, or ask questions through correspondence, an inefficient method. Here Davis and Watts could have helped. The firm not only had technical expertise, but as a manufacturer of Bell equipment, it often received word of important innovations and improvements before other agents.[37] Making Davis and Watts solely responsible for the design and production of southern telephones and equipment would have increased the firm's ability to supply the South with needed technical aid.

There is little doubt that the Baltimore firm could have met these technical needs. It already had a manufacturing contract with Bell that allowed it to produce and modify telephone equipment—though not telephones themselves—and had performed impressively.[38] But the manufacturer was not yet prepared to meet Ormes's demands for capital. Though Davis and Watts had furnished the switchboard for Richmond on sixty days credit, it made it clear that it could not continue to make such advances without steady returns.[39] The key to implementing Ormes's plan was to provide Davis and Watts with some steady source of profits from which the company could supply needed short-term capital.

Telephone production promised to fill this need. Telephones were easy to make and profitable to sell, and they comprised the bulk of manufacturing output in the industry. But Bell managers refused to license Davis and Watts to make telephones from their patents, though at one time it seems they suggested to Ormes that they might do so.[40] Theodore Vail, however, quashed this hope and kept production of telephones in Boston. Afraid of losing hold of the valuable telephone patents—the most important property Bell possessed—he only allowed Davis and Watts to make supplemental equipment.[41] Vail refused to modify this restrictive policy, preventing Ormes from making similar arrangements with any other company.[42] Without the right to make telephones, no manufacturer could meet the southern agent's needs.

While unsuccessful, Ormes's plan clearly fit into his evolving entrepreneurial strategy for the South. Unlike his fellow southern agents, Ormes had recognized that the future of telephony depended on new sources of capital and technical expertise, and the swift adoption of new equipment. Not constrained by the tenets of localism, Ormes also sought contacts with parties outside of the South to meet these needs. Undeterred by his initial failure with Davis and Watts, the agent continued to search for resources, finally succeeding in establishing contact with another outside firm—his rival, Western Union.

In the summer of 1879, Ormes joined forces with DeLancey Louderback, a representative of the telegraph corporation's subsidiary, Western Electric. Louderback hailed from Davenport, Iowa, and had come East in the early 1870s to manage a Western Union office in Philadelphia. Like Ormes he possessed an expansive entrepreneurial vision and the ceaseless energy of a professional promoter. Restlessly seeking opportunities, he left his employer and opened independent telegraph offices in Philadelphia, New York, and Washington, which he sold at a profit to Western Union in 1872. Returning to the Midwest, he got a job with the A&P Telegraph Company, Western Union's chief competitor, and then later returned to Western Union in Chicago. Not content with a management position, he opted for the job of eastern sales agent for Western Electric. Still unsatisfied, Louderback continued to find outlets for his promotional talents by aiding Western Union president Anson Stager in public utilities ventures.[43]

In Louderback, Ormes found a partner who matched his own ambition and spirit. Meeting through Daniel Carson, Ormes and Louderback first came together while still officially working for competing organizations. Despite the supposed competition between Bell and Western Union, the pull to cooperate was too strong. Louderback, more interested in exploiting opportunities than conforming to company policy, sold the Bell agent electrical equipment on sixty days credit.[44] During the next few months, their contacts grew. With help from Louderback, and several other men, mostly nonsoutherners, Ormes secured district and exchange contracts in a string of southern cities. With S. H. Fishblate and F. W. Foster, Ormes executed a Bell contract for Wilmington, North Carolina; with George Mason, D. J. Parker, and W. D. Wickersham, he obtained one for the entire state of Alabama. With Louderback, he planned to open exchanges in Mobile, Selma, and Eufala, Alabama; Jacksonville, Florida; Macon, Georgia; and Raleigh, Charlotte, and Goldsborough, North Carolina.[45] Timely help from Louderback was enabling Ormes to begin to realize his regional strategy.

These achievements, though important, were still threatened by the lack of capital. Without funds, the contracts remained just pieces of paper. Bell assisted Ormes directly at this juncture, remitting rentals on Richmond telephones in exchange for stock in the exchange company. This policy foreshadowed Bell's later move into ownership of regional telephone companies.[46] More importantly, Bell and Western Union began negotiating to settle their differences and provide the resources the telephone needed.[47] This important accord was struck first in the South, where Ormes and Louderback arranged to bring Western Union's vast resources into Bell's southern telephone operation.

Though cooperation offered benefits to all parties, the negotiations leading up to this settlement were fraught with tension and conflict. Each of the parties had different interests. Ormes feared that a settlement between Western Union and Bell would squeeze him out of the South.[48] Despite the small stature of his enterprise and his shortage of capital, however, Ormes was actually in good position. He alone had the intimate knowledge of the southern market necessary for further expansion. Perhaps more importantly, he had the sort of strong yet flexible personality needed to bring rivals together. Bell management itself feared their agent would sign a separate peace with Western Union that would tip the competitive scales in the telegraph firm's favor. Western Union, though the largest actor, was besieged with its own problems.

Since 1879, the telegraph corporation had been under attack from Jay Gould's American Union Telegraph Company, which sought to take controlling interest in Western Union from William Vanderbilt. In the North, American had engaged in a successful flanking operation that was driving down the price of Western Union stock. Gould offered to purchase Bell's exchanges and deploy them in his war with Western Union.[49] In July, the infamous financier tried similar tactics in the South, offering to buy out Ormes.[50] This threat provided a needed spur to bring Western Union to the bargaining table with Bell and Ormes.

In the hot summer months of 1879, a handful of men hammered out the agreement that determined the shape of the southern telephone industry, and that of the rest of the nation, for the next fourteen years. No one involved could have said for sure where things would end up. Bell had the choice of settling with either telegraph company—Western Union or American Union. Ormes had perhaps the clearest position. With an offer from Gould already on the table and one from Western Union expected, he wrote "it is purely a matter of dollars" which company he went with. Though he disliked the American Union Company, it was "about even chances," which he would choose.[51] Similar thoughts were running through Vail's mind, and he advised Ormes that "*the fence* is the best place to sit on just now."[52]

Still, neither Ormes nor Vail could have been totally confident positioned between two titans like Western Union and Gould. By August, fence sitting was getting to be uncomfortable. "It is fearful business this waiting so," Ormes admitted, "but it is a 'life affair,' and *must be done*."[53] The difficulty of Ormes's position increased considerably as the season passed its meridian. By mid August, Western Union was emerging as the clear choice for the Bell parties, but Ormes fell prey to debilitating hayfever in the summer heat.[54] The heat of bargaining also remolded several crucial relationships. Ormes strengthened his ties with Theodore Vail, whom he termed "the only *sane* man or real *square* one in the lot."[55] But the agent grew alienated from his old allies Gardiner Hubbard and James Tracy, whom he found "too *hoggish* to live in houses." Moving away from Gould, Ormes was drawn much closer to Western Union, convinced that "their policy is far the more liberal and . . . honorable."[56] Finally, at a dramatically late hour Ormes reported to Carson in Virginia that, "after two days full of hard work on the part of Vail and I and Louderback, we have got Mr. Forbes in his shirt sleeves in the act of writing out a contract that he will sign."[57] Rejecting Jay Gould's offer, Ormes signed a tripartite agreement with Bell and Western Union that ended telephone competition in the South.

The price of settlement had taken its toll physically and psychologically on Ormes. "I am almost dead," he wrote to Carson in the final days of negotiation.[58] But the tripartite accord conceded to Bell, and to Ormes as the company's agent, a free hand in the southern telephone market. In exchange for this valuable right, Bell agreed not to engage in a message-for-hire business. As understood by Western Union, this stipulation prohibited the telephone firm from providing intercity telephone service, leaving the profitable long-distance communications business in its hands. As specified in the agreement, moreover, the parties formed a new company, with Ormes and Western Union dividing the stock. The telegraph corporation received 50 percent ownership, Ormes 37.5 percent, and Louderback 12.5 percent in what was soon styled the Southern Bell Telephone and Telegraph Company. As Louderback was an agent of Western Union, his shares were held in trust by that company, giving it clear majority ownership. Ormes and Louderback then turned over their exchange contracts to Southern Bell, giving it sole responsibility for the telephone business of the South.

To solidify relations between the telephone and telegraph firms, the agreement also set up a system of reciprocal message exchanges. Southern Bell passed on intercity communications to Western Union for transmission over its telegraph wires; Western Union did the same with messages for local distribution in towns where Southern Bell had an exchange. This service never amounted to a large percentage of the business, however, and did not create the joint telegraph-telephone communications system that Ormes and Western Union may have envisioned.[59]

Bell itself received no stock in the new southern firm, but it gained an open field in the South for its telephones. The tripartite agreement also served as the basis for a similar national contract between Bell and Western Union, which put an end to telephone competition in the United States until Bell's patents expired.[60] The national settlement clearly would have come about anyway; Western Union was under too much pressure from Jay Gould to keep its resources dedicated to the fight with Bell. And as the corporation's president, Anson Stager, had already perceived, the telephone would make "immense" inroads on local telegraph operations.[61] Nonetheless, the settlement in the South contributed to the national agreement. With resources for the telephone scarce in the South, both Ormes and his adversary benefitted more by cooperation than competition, and thus were able to come to terms quickly. Located away from the main Bell–Western Union fray, the region served as a testing ground for the national settlement.

For Ormes, alliance with Western Union provided the resources he had so long sought for his business. Southern Bell opened for the South two new channels of outside capital—Boston and Chicago. Western Union provided much of the early money for SBT&T. But a group of Boston financiers, many probably early backers of Ormes's agency, also took an interest in the new firm. Their participation was especially important in the first few years. In 1880, Southern Bell had assets consisting of eleven exchanges serving 1,246 customers, coverage still below what one would expect in its vast territory.[62] Accordingly, the new firm planned to open nine more exchanges quickly, and it needed funds.[63] Southern Bell also had other claims on its capital. The company had already issued $150,000 in stock to pay the franchise claims of the tripartite partners.[64] It also gave $10,000 in stock to Richardson and Barnard for their Tybee Telephone Company, which operated the Savannah exchange, and bought out the Richmond and Norfolk exchanges for $32,000 and $9,000, respectively. In addition, James Ormes received $20,000 in stock for his private line franchise.[65]

To finance expansion and fund these initial expenses, Southern Bell issued $137,000 in new stock its first year, most of it to a group of financiers and manufacturing firms from Boston. Of the 3,401 shares of the company's stock traded the first year, 63 percent involved this Boston group. Since some of this trading was in franchise stock, the Bostonians' contribution to paid-in capital may have been even greater.[66] Only 5 percent of the stock traded involved parties from New York, Wisconsin, and other northern states other than Massachusetts. The South's contribution was 3.2 percent. While these figures cannot be taken as anything more than a rough approximation, it is clear that the majority of new finance during the first year of Southern Bell's existence came from a small band of Boston investors. By 1881 they owned 33 percent of the firm.[67] Indigenous southern capital contributed very little to the enterprise. The southern telephone industry was almost wholly dependent on outside resources.

The Boston capitalists continued to supply injections of new capital in the following years as well. Southern Bell issued $48,000 in stock in 1882 and $34,000 in 1883.[68] Two Boston firms, H. C. Wainwright and Company and Jackson & Curtis, provided between them $10,000 each year.[69] Western Union, as majority owner, was obliged to take the largest share both years—$29,709 in 1882 and $19,800 in 1883. The balance was supplied by scattered small investors from New York, the South, and the Southwest. While quantitatively smaller than Western Union's contribution, the Boston group's financial involvement was vital to Southern Bell. The telephone company was fairly closely held, and its stock did not command immediate public recognition. The Bostonians, trading SBT&T shares among themselves, created a lively informal market for the company's securities.[70] This trading encouraged investors to make equity purchases, since they could be sure of finding buyers for their holdings if they chose to sell later. Southern Bell could therefore rely on long-term funds and reduce immediate outlays for interest payments on short-term debts.

Financially, then, Southern Bell was something old and something new. As was typical of firms in the nineteenth century, working capital was scarce and business had to be done on a discount basis with suppliers.[71] Employees and outside creditors frequently received payments in the form of claims on future earnings. Investors were drawn from a narrow field—friends of the founding entrepreneur or manufacturing concerns willing to speculate in industrial stock before the rise of a formal market for such securities.[72] But Southern Bell also represented an important innovation for the South. Incorporated in New York, it was a large organization capable of attracting interregional flows of funds without the assistance of a financial intermediary. Tied to what was still the largest corporation in the nation, the telephone company was quite an unusual presence in the backward southern economy.

Managerially as well, Southern Bell differed significantly from other southern firms. Its executives came from the North and Midwest through its two important interfirm contacts, Bell and Western Union. The Chicago corporation provided most of SBT&T's managerial staff during its first five years of life. Maintaining tight control over its holdings of the southern company, Western Union distributed shares to members of its New York branch. These men, John van Horne, Anson Stager, the first SBT&T president, Augustus Schell, Norvin Green, Henry Plant, and Edwin Morgan, formed Southern Bell's executive committee between 1880 and 1885.[73] After 1881 Jay and George Gould's names were added to this roster, as the financiers had completed their takeover of the telegraph giant. Western Union leadership did not go unchallenged, however, as the important Boston group refused to be denied a voice in Southern Bell. They allied with the parent Bell organization, which after 1885 also looked to take greater control of its southern licensee.

Now styled American Bell, the parent telephone company suddenly realized that its hold on the vast southern territory was shaky.[74] Bell–Western Union relations deteriorated quickly after Gould's ascendance. Gould, realizing he had been shut out of the telephone market by the 1880 settlement, resorted to his usual tactic, filing suit in Massachusetts court to void the contract. This move touched off two decades of bitter legal struggle, and helped to motivate Bell to gain majority interest in all of its regional licensees. In the South, however, the telephone firm had no means of obtaining ownership. Bell's usual method was to demand stock in exchange for a perpetual license to lease Bell telephones. But Southern Bell operated under Ormes's original ten-year license, which would not expire until 1889. With Jay Gould to contend with, the parent firm could not wait that long.

To increase its influence in the South, Bell maintained close contact with the Boston manufacturing companies, Jackson & Curtis and Wainwright, who had significant minority interests in Southern Bell. Through them it purchased 25 percent ownership in SBT&T by 1885. Bell also obtained stock from individual owners such as Charles Hubbard, Gardiner Hubbard's brother and an early investor.[75] Buying up these odds and ends, the telephone company was able to obtain a large minority interest in SBT&T, though not enough to wrest control from Western Union. Only in 1889, with the expiration of Ormes's franchise, did it have the leverage to become majority owner. After that date American Bell was able to force Southern Bell to relinquish 35 percent more of its stock in payment for a permanent contract, making it the dominant voice in the southern company.[76]

By the end of 1889, the southern telephone industry had completed a major transformation. Changes beginning with Ormes's agreement with Western Union and ending with American Bell's acquisition of Southern Bell had swept aside the old agency structure and consolidated the business under a new, large-scale firm. This reorganization put an end to the conflicts between Bell and southern entrepreneurs which had retarded early telephone growth. It also solved the long standing problem of resource shortages by bringing new capital and new entrepreneurial talent into the South.

During this same period, similar changes were taking place in telephony throughout the nation. Elsewhere, however, the pattern of change was significantly different than in the South. In New England, the old system of general and subagents also gave way to consolidated operating companies. But that region, lacking the debilitating conflicts over entrepreneurial strategy that blocked growth in the South, drew on its indigenous business population to bring about the transformation. With its rich regional supplies of capital, technical skill, and entrepreneurial talent, New England accomplished on its own what had to be induced from the outside in the South.

The reasons for this difference go back to the strong interest that northern entrepreneurs exhibited in modern industries such as the telephone. Just as they were prepared to embrace the new technology of the exchange, they were ready to create larger firms to consolidate small local exchange companies and agencies.[77] The men who carried out this second step of the process were sometimes original Bell agents, but they were often sharp speculators similar to those consolidating New England's railroads at the time. Their actions seem for the most part to have been beneficial to telephone growth, and they did not seek quick profits through stock jobbing and fraud. Nonetheless, they were motivated by the pecuniary rewards of corporate promotion and securities trading, which could be considerable in the early days of big business.

Some disputes between these new entrepreneurs and Bell still arose, but of a very different sort than those which plagued the South. Northern promoters were not timid when it came to large investments in extended regional companies, as were southerners. Nor were they inhibited by restrictive ties to a local economy. Rather, like James Ormes, they were professional promoters in search of profits from organization. The disputes that did arise in the North came when these men aggressively consolidated the telephone business and threatened to leave Bell without a role to play. Such conflicts contrasted significantly with those which arose in Atlanta, where the timidness of indigenous entrepreneurs was the issue.

The most important of the New England consolidators was the Lowell Syndicate. Headed by Charles Glidden, William Ingraham, and L. W. Downs, this group tried to reap windfall rewards by taking over numerous small firms in western Massachusetts, Vermont, Maine, and New Hampshire.[78] Their sharp financial dealings made good Gilded Age newscopy, inflating their reputation as telephone moguls. The leader of the triumvirate, Glidden, was a self-styled financial maverick who combined speculative panache with hard-headed business sense. Later, he achieved fame by the stunt of riding an automobile around the world and through his barnstorming adventures in the air mail delivery service. But Glidden and his comrades had substance as well as style.

The Lowell Syndicate originated in 1879 when Glidden, still a Western Union operative, convinced William Ingraham, a well-to-do grocer, to buy the newly opened Lowell telephone exchange. This property was owned by the old New England Bell Company, a subsidiary started by the main Bell Telephone Company to supervise regional promotion in the valuable New England territory. Organized prematurely and continually short of capital, New England Bell gladly sold its holding in Lowell to the partners.[79] Glidden and Ingraham borrowed $6,000 from a local bank, and sold another $3,000 in stock to finance the purchase.

While this exchange was the first part of what would become a large portfolio of telephone properties, the Lowell partners did not spend all of their funds on acquisitions. Though they eventually realized a considerable profit from stock purchases and consolidations, they were innovators as well as jobbers. With Ingraham's open purse—in contrast to the tight-fisted policies of New England Bell—Charles Glidden built two early long-distance lines. One stretched from Lowell to Boston, and the other from Boston to Granitville. In the Boston-Lowell region manufacturers clamored for lines from their outlying factories to the cities. In the period of competition with Western Union, the company that served them first received their patronage both as customer and as investor. Glidden successfully met this demand, getting Theodore Vail to agree to a strategic rate decrease on the long lines, despite the parent firm's rather rigid pricing policy.

Though Glidden had to fight with Bell to achieve his ends, by March 1879 he reported, "all the principle houses and establishments are connected with our system here [Lowell]."[80] His ability to change Bell policy was crucial at this early stage. The parent company had yet to take full responsibility for telephone development, and innovations from the field such as Glidden's were still quite important. Glidden also recognized the value of long-distance connections to the growth of a regional telephone industry, noting how Boston's central position encouraged the construction of lines and exchanges in surrounding cities such as Lynn and Lowell. The interrelated nature of local and long-distance service prompted him to build and organize well-financed firms to manage the individual exchanges in the region. Bell itself would eventually follow the same policy, but in 1879 the idea was being promoted by Glidden, who, working in the field, was more intimately familiar with the demands of the region.

In accord with his perceptions, Glidden in April of 1879 branched out from Boston and Lowell to nearby towns. He added Lawrence and Haverhill to the Boston-Lowell circuit.[81] Using his powers of persuasion, the telephone promoter also began to merge small exchanges in New England into the Boston and Northern Telephone Company.[82] This organization encompassed the towns of Concord and Littleton, New Hampshire, St. Johnsbury, Vermont, and the county of Essex, Massachusetts. In 1882, Glidden and his partners took over the Newburyport exchange as well. Throughout New England, they acquired, consolidated, and refunded the small exchange companies that had sprung up between 1878 and 1880. Linking many of them with short-haul toll lines, the entrepreneurs laid the basis for a regional telephone network. By 1882, the Lowell Syndicate controlled 19,680 telephones and part or all of six companies. Its operations stretched west to New York state and north through Maine.[83]

Other New England businessmen also joined in the task of reorganizing their region's telephone industry. Charles Cutler and Joel C. Clark formed a company that operated in central Massachusetts in a fashion similar to the Lowell organization. In 1879 Clark, a late applicant to Bell for an agency, joined forces with Cutler, a wholesale grain and flour merchant whom he had met through his library society. With support from Cutler and funds from his own printing company, Clark started the Central Massachusetts Telephone Company of Framingham in November of that year. The firm promoted telephones in small New England towns, where urban financiers were reluctant to advance capital for the business. Clark and Cutler also saw that profits could be made by centralizing small, unconnected telephone operations. Between 1880 and 1881 they acquired the Hampden and the Bay State Telephone companies. These purchases at first pressed their modest resources, but eventually the partners earned sufficient returns to attract new capital. With the help of other investors, they reorganized their operations in 1881 as the Massachusetts Telephone Company, and the next year they sold out to the Lowell Syndicate.[84]

The Lowell partners and smaller entrepreneurs such as Clark and Cutler responded to local conditions in their operations, but with a wider vision, one consistent with that of the Bell Company.[85] Clark, like Glidden, recognized that long-distance service would be valuable to local firms and built feeder lines from his exchanges to the Boston-Worcester trunk line.[86] He knew that local exchanges had to become part of a larger telephone network. After his enterprise was absorbed by the Lowell Syndicate, Clark wrote that such consolidation was simply a necessary part of industrial progress.[87] In retrospect he viewed his own efforts as merely the first step in a process that culminated in an integrated telephone system controlled by a single firm—American Bell.[88]

American Bell itself became involved in the process of consolidation after 1880 through its regional subsidiary, New England Telephone and Telegraph (NET&T).[89] Like Southern Bell, this operating company was formed following the Western Union settlement, though it was controlled by Bell interests in Boston.[90] NET&T ran the important Boston exchange, but it did not dominate its territory in the same way that Southern Bell controlled the South. Other promoters, the Lowell group most notably, still directed much of New England's telephone business.

This division of the market proceeded without conflict for a short while. Indigenous New England entrepreneurs worked together with Bell-controlled NET&T to create a rough sort of organizational hierarchy in the industry. At the top of this structure sat American Bell, which was the preeminent player due to its ownership of the telephone patents. Larger firms such as NET&T and the Lowell Syndicate controlled major cities and maintained strong links to the parent organization. In suburban areas, entrepreneurs such as Clark and Cutler built smaller firms. Below them, in still smaller, more remote towns such as Brockton, Maine, and Burlington, Vermont, local, Bell-licensed exchange companies sprang up. Some of these were absorbed by the larger firms, but others held out. Linking all of these levels together were informal mechanisms of cooperation and control.[91] The lines of authority in this structure were complex, but they worked fairly well, in large part because all parties shared in a basic consensus over the nature of telephone development. Local entrepreneurs and Bell executives both agreed that some sort of interconnected system operating over a broad territory was the most profitable way to run the business.

Other forces, however, soon brought the two largest organizations—NET&T and Lowell—into conflict. Despite wide participation from the New England business community in the telephone industry, some promoters achieved greater market share than others. Few people at this time were able to mobilize capital for a venture like telephony, and those who could gained a distinct advantage over the others. Adept promoters such as the Lowell partners and the New England Company earned high profits buying and trading telephone company stock, encouraging them to make more acquisitions. The financial advantages of large-scale organization pushed consolidation forward, causing NET&T and the Lowell group to absorb more and more firms. Eventually, the territory became too small for two giants.

The Lowell Syndicate and Bell, in agreement over the basic strategy of telephone development, clashed over the pecuniary rewards of growth. American Bell had shifted from a patent-holding association to a managerial firm. It earned relatively less of its income from royalties and more from the profits of ownership in regional telephone firms. This shift brought it in direct competition with the Lowell group, whose aggressive policy of acquisition was depriving Bell of profits. Other considerations may have also pushed the parties into conflict. The success of Southern Bell must have impressed upon Bell managers the advantages of a single organization operating in a wide territory. By the same token, the conflicts Ormes had experienced with indigenous entrepreneurs in the South may have made Bell wary of relying too heavily on local businessmen. New England had as yet experienced none of the problems that plagued the South, but the danger was always present. Bell had only recently eliminated independent-minded agents from its organization. The Lowell group, which was making substantial headway in New England, might itself switch from a useful ally to a dangerous competitor. Beginning in 1883, therefore, American Bell moved to take control of the Lowell Syndicate's properties. After an acrimonious struggle, the company finally succeeded in convincing Charles Glidden to sell his share, over his partners' objections, thereby allowing Bell to obtain controlling interest.[92]

By the late 1880s, New England, like the South, had in place a regional operating company structure firmly under Bell control. One can find in the two regions similarities that led to this development. In both places the needs of the new technology of the exchange and the inadequacy of the old agency system were powerful inducements to change. The agency structure especially had proved itself a hindrance to innovation and change. As agents acquired knowledge and skill, learned to use new equipment, and gained experience in their work, they grew protective of their territory. Having made a substantial investment of personal labor and risk-taking in their franchises, they were unsure of what return they could expect, or even what an adequate return for their efforts would be.[93] Under these conditions, they grew reluctant to make further investments, necessitating the rise of a new type of organization—the regional operating company.

Overall, however, comparisons between New England and the South reveal striking differences in process and result. In New England, local entrepreneurs themselves brought about needed reorganization. As had been true since the start of telephony, local supplies of capital, entrepreneurship, and business skill were abundant and available for telephone promotion in New England. Under these conditions, reorganization proceeded smoothly, as the Bell corporation and local parties shared a common view of what telephony required. Those conflicts that arose came mainly out of Bell's growing interest in taking more of the profits of telephone promotion, not out of an entrepreneurial conflict between Bell managers and local businessmen. By contrast, change in the South came about because of the concerted efforts of three outside parties—James Ormes, the Bell Company, and Western Union.[94]

Given the nature of business organization in the South at this time, one might have expected the southern telephone industry to have been divided among numerous small regional and local firms. It should have been less, not more centralized than New England's. Because Ormes and the managers of Bell had innovative ideas, however, they needed a special type of organization like Southern Bell.[95] Going beyond the restricted, localistic outlook of southern businessmen, they had already conceived of the idea of a telephone system. A shortage of capital and manpower, as well as resistance from southern entrepreneurs to these plans forced Ormes to form a large-scale organization funded and staffed from outside of the South. As a result, the South had one of the most centralized structures in the telephone industry. New England, where local participation was more active and effective, actually maintained a number of smaller firms that continued to operate beside New England Telephone and Telegraph.[96]

Southern Bell provided Ormes with the means to overcome the early problems of promoting the telephone in the South. Tied closely to Bell and Western Union, this organization compensated for the limitations of southern entrepreneurship and mobilized outside capital in a way that small southern firms or individual entrepreneurs on their own could never have.[97] Able to circumvent the resistance of men like the Browns, who continued to view the telephone from a parochial, local perspective, the company institutionalized the new entrepreneurial strategy Ormes had introduced into the South. Perhaps the clearest mark of its success was the desire of local southern investors who had refused to invest their capital in the southern telephone industry earlier to subscribe to the company's stock.[98]

The early history of the southern telephone industry also reveals the importance of creative individuals in the process of economic development and technological diffusion. Without the vigorous participation of a James Ormes, the South might have missed out on the telephone altogether. Ormes possessed the skills needed to operate effectively within southern culture, but he also had the broader entrepreneurial outlook conducive to modern industry. Positioned between the worlds of the advanced North and backward South, he fit the technology of the one to the markets of the other.

Previous experience played an important part in Ormes's success. Having come from the North, he was familiar with conditions in that more advanced sector of the American economy. He maintained close association with this world during his time in the South, and drew on contacts such as Daniel Carson and DeLancey Louderback he had made earlier. Ormes's earlier experiences as an agent for New England mining investors operating in the Far West also gave him some familiarity with regional business cultures away from the urban East. He appears to have been one of that large group of mostly anonymous independent promoters who came on the American business scene after the Civil War. Before the creation of large managerial structures, these men spread modern industry from its roots in the industrial East and Midwest to the remote corners of the nation. Tied to no one region, possessing an unshakable belief in the value of formal organization, thoroughly convinced of the benefits of modern industry, they were the vanguard of the national organizations that would emerge with increasing frequency as the nineteenth century drew to a close.

Apparently Ormes himself recognized his own special role, for he quickly left the South after his task was accomplished and sought new opportunities in even more distant lands. Departing for Europe in 1880 on account of his health, he took a position with Bell's Oriental Telephone Company, where he attempted to rent telephones in the Far and Middle East. In what was surely one of the most far-flung telephone agencies of the time, Ormes, using Daniel Carson in Virginia as his purchasing agent, shipped telephones to operatives in Hawaii, Australia, India, and Egypt, while aiding Bell promotional efforts in Texas and Mexico.[99]

Ormes's entrepreneurship grew out of his own perceptions of and values regarding business development. From the very beginning the Bell agent embraced the notion of a telephone system. Bringing about system development, Ormes understood, required an orderly and systematic approach to telephone promotion that encompassed the entire southern territory, not simply one locality. This view of telephony cannot be traced to existing economic and technological conditions in the industry. Before 1885, long-distance service and the linking up of individual exchanges was still quite limited, particularly over distances exceeding a few miles. Systems were already apparent in railroading and telegraphy, however, and most astute telephone men, such as Theodore Vail, Thomas Doolittle, Charles Glidden, and Ormes, expected they would emerge in telephony as well. For Ormes, the value of a system was an article of faith that, like his profound religious faith, guided him in strange and hostile territory.

So strong were Ormes's basic business precepts that they never left him. Following his successes in the South, he became involved with John Pender, the "king of cables," who had followed Cyrus Field in building a successful transatlantic telegraph cable. With Pender Ormes developed, though never carried out, a plan for a worldwide system of interconnected cables for transcontinental communications. The man's fascination with systems did not escape him even while he was on vacation. While in Venice he wrote for his children a series of sketches of the city. The theme: Venice attained its glory through the counsel of rational men who overcame the obscurantism of the Church and built their city's canals, waterways, streets, and sewers.[100]

Calculations of rates of return played little part in this entrepreneurial vision. Though Ormes sought profits, his pecuniary motivations were of a more general sort. Believing that the telephone would be an important new product in the South, he was content to wait years before he saw a reward. Though Ormes did make money on his labors, he also defined success in broader terms. By forming an early connection with a new industry, he gained important contacts, knowledge, and opportunities that he later exploited. This sort of motivation was crucial to Ormes's innovative activities. Since he had to make a market for the telephone, ex ante calculations of profit had little meaning. More important was his belief that money and other rewards would flow from successful entrepreneurship.

Perception, vision, and creativity of this sort are important aspects of entrepreneurship; without them, the entrepreneur can do little more than respond to existing environmental conditions. At the beginning stages of an industry, conditions can change rapidly and may seem to point in several different directions. One way of defining entrepreneurship is the ability to seize on the right course before it becomes obvious. An even bolder definition would be that the entrepreneur is one who is able to shape and influence the direction of change. In the case of the telephone, for example, was the concept of the system "embedded" in the technology, or did the creative efforts of Ormes and others push the technology of telecommunications in this direction? The evidence available provides no clear answer. Southern entrepreneurs offered no alternative to the path marked out by the system-builders of the Bell Company. That company, moreover, held the key patents on telephone technology, enabling it to shape the technology as it desired. There is no way of knowing what other lines of development might have been possible had conditions in the South been different or had the industry been more open and competitive. As we shall see, however, different conditions later did bring forth different approaches to telephony in the South.

In the short run, the reorganization of the telephone industry between 1878 and 1880 brought clear benefits to the South. It helped to overcome resistance to change and innovation that would have severely retarded the development of an important new industry. Neither Ormes nor the other men involved in this effort acted out of altruism; the purpose of Bell policy was, as Vail expressed it, to, "protect to the fullest possible extent the present and future interests of our company, without restricting by unnecessary conditions the fullest possible use of the telephone."[101] Increasingly, these two goals would become incompatible, especially in the South. For southern consumers and businessmen, the "fullest possible use" of telephones required local control, and the presence of a large corporation funded and directed from the North struck many as economic colonialism. Conflicts between southern interests and the Bell organization continued, and the differences between the southern and northern entrepreneurial cultures reappeared again and again as Bell built its system.

Chapter 4

Rapid Change and Technological Conflict

The need for further reorganization of the telephone industry in the South was not immediately apparent in 1880. Southern Bell had overcome the limitations of southern entrepreneurship that had inhibited reception of the telephone in the region, and with financial support from Western Union, the company was able to open forty-two exchanges during the following nine years, adding 5,127 customers. Though these accomplishments were less impressive than those of Bell companies elsewhere, they did demonstrate that Southern Bell had tapped into the latent telephone demand of the South. Nonetheless, the 1880s were rife with conflict between American Bell and its southern licensee.

In 1880 the parent organization sat atop what was a fairly stable industry. Southern Bell was one of a number of regional telephone operating companies serving various markets in the nation. ABT had acquired partial control of most of these companies, providing needed coordination in the industry. Although its monopoly was secure, the firm did not stagnate. Though it maintained relatively high prices, earned monopoly profits, and restricted entry into its market, it also sought to develop new products and services, the most important of which was long-distance service. It was in this segment of the industry that conflicts with the South emerged.

Communications between exchanges had been tried as early as 1878, but the formal beginnings of organized long-distance service came in 1885 with the founding of American Telephone and Telegraph Company (AT&T) in New York. Chartered originally as the long-distance subsidiary of the entire Bell organization, AT&T's first tasks involved rather modest experiments with intercity lines, mainly in the densely packed New York - New England region. Very quickly, however, long-distance service became the driving force of telephony. Protected by an unbreachable patent monopoly, AT&T engineers were able to formulate ambitious plans to make the entire nation one interconnected system. By the end of the century, the former long-distance subsidiary would control the entire Bell empire.

To build its system, AT&T had to standardize the technology of telecommunications. The system was to be physically interconnected, and thus problems, breakdowns, and inadequate service at one point could affect service elsewhere. The quality of equipment used by operating companies became an overriding concern from the first stages.[1] Poor quality lines and equipment at the local level would create unbalanced circuits and increase induction noise, restricting long-distance transmission and reducing quality. Only if all regional companies used compatible technology and followed standard procedures was communication between exchanges feasible. Bell also needed regional feeder lines to its long-distance trunks, to provide adequate demand for its system and make it profitable. This work had to be carried out by regional firms, but they had to follow nationally set standards in doing it.

While the parent organization was moving toward standardization, managers of several regional operating companies, Southern Bell included, were proceeding in another direction. Southern Bell executives found that they had to respond to local conditions and the needs of southern customers if their firm was to make a profit. These conditions and needs dissuaded them from adopting the latest innovations in telephony. Between 1880 and 1885, Southern Bell made considerable expenditure on construction, opening new exchanges and reconstructing old ones.[2] After 1885, however, the southern market for telephones sagged, and the company was forced to close more exchanges than it opened, though it did add a net total of 3,000 customers to those exchanges that remained (see table 4.1). With demand uncertain at best,

Table 4.1. New Customers and Exchanges, Southern Bell

Year	New Exchanges (net)	New Customers
1880–81	9	251
1881–82	15	197
1882–83	7	899
1883–84	8	866
1884–85	4	445
1885–86	–5	421
1886–87	0	575
1887–88	0	698
1888–89	4	775
1889–90	0	859

Source: SBT&T *Annual Reports*, 1880–90.

Southern Bell began to restrict its long-term outlays, and construction as a percentage of total expenditures fell (see table 4.2). The company made steady profits, but it reserved little for depreciation and made no major investments in new equipment or reconstruction, preventing it from building the lines and adopting the type of equipment compatible with the emerging long-distance network. The result was the formation of a technological style at Southern Bell that conflicted sharply with AT&T designs.[3]

To AT&T engineers, the failure of Southern Bell to adopt new equipment reflected the narrow outlook of the company's managers. Yet the fault was not all with the operating company. American Bell offered its licensees little incentive to follow its plans, as most profits from the long-distance business accrued to the parent organization. AT&T's bold new undertakings also ran against the obligations Southern Bell had to fulfill to its owner, Western Union. Western Union had aided Southern Bell in the beginning by serving as the firm's main source of long-term capital. It financed short-term expenditures as well through cash advances. These infusions of funds had enabled the company to invest in an important new type of central office equipment—the Law Telephone Switchboard. But Western Union's financial policies soon changed. The

Table 4.2. REVENUE AND EXPENDITURES, SOUTHERN BELL (IN DOLLARS)

Year	Revenue	Total Expenditures	Construction Expenditures	Operating Expenditures	Percent
1880–81	90,409	65,761	—	—	—
1881–82	126,872	84,267	74,397	9,870	88
1882–83	175,152	119,230	46,263	72,968	38.8
1883–84	241,381	188,651	72,309	116,342	38
1884–85	279,283	218,112	41,542	176,570	19
1885–86	312,965	244,464	17,628	226,836	7.2
1886–87	348,519	254,501	34,433	220,068	13
1887–88	396,354	278,354	43,902	234,452	15
1888–89	430,798	313,421	27,960	285,461	8.9
1889–90	376,820	492,319	45,893	446,426	9.3

Source: SBT&T *Annual Reports*, 1880–90.

Note: Percent = construction as a percentage of total expenditures.

Chicago corporation supplied no injections of new capital after the first few years and refused to increase Southern Bell's debt or seek new sources of funds.[4] At the same time, it began to pay itself substantial dividends, averaging 11 percent per year between 1880 and 1889.[5] These policies drained Southern Bell's financial resources at a time when it needed them to undertake new work. Until late 1889, when American Bell finally assumed majority ownership of its southern licensee, there was little that could be done.

The structure of southern capital markets provided no relief from Western Union's fiscal conservatism. In New England, highly profitable urban exchanges attracted investment from the New York and Boston money centers. These flows of funds merged with local capital, providing sufficient capital to spread the telephone to the hinterland. But the national capital market, a relatively recent development, did not yet reach into the remote South. As a result, Southern Bell had little access to outside capital, except through large firms such as Western Union and American Bell. By 1885 almost all of the southern company's funds came from these interfirm sources, as the early Boston backers of Southern Bell were replaced by ABT. This situation, though common in new industries in the nineteenth century, did not provide sufficient funds to carry Southern Bell into the new era of telephony.

Southern Bell was clearly caught in a bind. It earned steady profits, but they were paid out as dividends. Western Union, happy with the income it was receiving, had no incentive to provide more capital, and Southern Bell was unable to seek funds on its own. Between 1885 and 1887 income from the private-line business fell, while demand for exchange service was uncertain, reinforcing Southern Bell's caution in expenditures.[6] Nor did American Bell, still short of capital, have money to lend its southern licensee. At the same time, however, AT&T was pressuring Southern Bell to undertake substantial construction to upgrade and expand its equipment in preparation for long-distance service.[7] Under these conditions, the immediate cost of new technologies became paramount. Restricting its long-term outlays, Southern Bell began to look with favor on low-cost equipment, against the wishes of the parent organization.

Southern Bell's choice of central office equipment reflected this concern with capital expenditures and regional conditions. In 1880, the company had been faced with the decision of what type of central office equipment it would employ in its new exchanges. It selected the switchboard of the Law Telegraph Company for its important operations in Atlanta and Richmond, and later installed this same equipment elsewhere. The Law Telephone Switchboard had been invented in 1878 by William Childs, who based his device on the same principle he had used for a telegraph switchboard.[8] The Law board was fast, easy to use, and relatively inexpensive. It quickly attracted the attention of D. H. Louderback, Ormes's partner, who expressed an interest in it as early as 1879.[9] By 1885 the switchboard was in use throughout much of the South.[10]

At first it seemed that Southern Bell had made a wise choice. In 1880 the Law board was at the forefront of telephone technology, competing with other equipment in speed, efficiency, and cost.[11] By 1882, however, ABT engineers began to express dissatisfaction with the device. At first, they complained that it violated patents held by Western Electric, the important manufacturing concern Bell had recently acquired. In 1887, Thomas

Lockwood raised other objections. He noted that the Law equipment did not fit well with new telephone technology such as metallic circuits necessary for long-distance transmissions.[12] Other Bell technicians doubted that the Law board could be expanded to serve large cities and areas of heavy telephone use.[13] Both Lockwood and Hammond Hayes reported to ABT president John Hudson that the Law system restricted the number of subscribers who could be connected in one place. Lockwood pointed out that the equipment limited efficient division of operators' labor, while Hayes raised the problem of designing multiple Law boards.[14]

Although these were serious objections to parent company personnel, they seemed of little significance to Southern Bell managers. The limit on switchboard capacity, for example, was not a major problem in the South, where demand density was still low.[15] It made little sense to SBT&T managers to make the investment necessary for a rapid changeover to a new system, when the benefits of such an investment lay in the distant future. Southern Bell did not need equipment for large exchanges, but a means of meeting the demands of small and mid-sized towns. Similarly, few southern customers made calls beyond their exchange area, obviating the need for equipment compatible with long-distance service. To Southern Bell personnel, operating in a territory where long-distance service was not yet a large business, the problems of the Law equipment seemed manageable, and the simple, inexpensive device continued to be used in the South.

It was not stubbornness or lack of vision that led SBT&T managers to take this view. Other places with similar needs also found the Law technology attractive. In St. Louis, George Durrant became an outspoken proponent of the equipment. Charles Wilson of Southern Bell wrote to George Phillips, president of the Central Union Telephone Company, a firm which served the rural Midwest, that the speed of the Law board was quite impressive and that other problems, such as connection with metallic circuits, could be overcome in time.[16] Southern Bell engineers themselves were already at work on some of these limitations. Phillips was impressed enough with the equipment that he believed these efforts should be continued and that Bell should not abandon it entirely.[17] While he recognized the problems of fitting the Law board to metallic circuits and the difficulty of using the Law system to record toll calls, he also noted that these were problems of concern mainly to AT&T, the long-distance company.

Problems with the Law technology clearly existed, though its supporters believed that they could be overcome in time. Had ABT continued to use the Law board and invested in it research time and money, it might have made the equipment compatible with its other technology.[18] But to the equipment's detractors at the parent company, its potential incompatibility with metallic circuits and other highly touted new technology such as the common battery telephone was too large an obstacle.[19] These problems indicated that the Law equipment would increase the difficulty of fitting together local and long-distance service, an issue that loomed large to managers and engineers at AT&T.

Such considerations, in fact, underlay all the parent company's objections to the Law board, even that it conflicted with Western Electric patents. In 1882, American Bell had acquired Western to manufacture telephones and equipment for its licensees. The manufacturer resented paying royalties to the Law Company to produce its equipment for Bell operating companies, however, for it had been in competition with this firm before the Bell acquisition.[20] ABT president Hudson was wary of such resentment, fearing that it would undermine the vertical structure of his organization. American Bell had bought Western Electric specifically to ensure that its regional licensees followed standard practices and procedures and employed standard equipment in the delivery of telephone service.[21] In this way it could enforce the policies necessary for an integrated system. Use of the Law board ran counter to these efforts at vertical control.

As the Bell system took shape, this point of conflict grew.[22] Control of technology was an important source of strength for Bell, and the firm did not like the idea of depending on an outsider for what might become a necessary piece of equipment.[23] Such dependency, Bell managers feared, would limit their ability to modify and shape their system of technology to suite their long-term strategy.[24] Just this sort of fear underlay Hammond Hayes's pleas that Southern Bell adopt a "new switchboard of the standard pattern," in Atlanta, rather than experiment with equipment of outside firms. Hayes noted how one "mistake" in a technological decision led to another, destroying all hope of standardization.[25]

Laying behind Hayes's statement was the contrasting technological mindsets of the members of Southern Bell and the parent company managers.[26] The thrust of ABT's technological style was to centralize the management of technology. The firm wanted to be able to plan from the top down, to direct telephone service as though it was provided by a single entity, rather than a variety of organizations. It sought to create a national network and needed tight control of operating company technology to do so. Most members of the southern firm did not share ABT's outlook, placing less emphasis on such goals. Parent company technicians condemned Southern Bell equipment as inefficient; but in fact they meant that it did not fit with their efforts to build a highly integrated national service. SBT&T managers, on the other hand, found their technology adequate to serve their market.

Many examples of what Bell engineers derided as southern technical inferiority were actually expressions of this conflict of technological styles. Though Southern Bell produced fewer innovations than other firms, those it did produce conformed closely to regional demands. The company's engineers tended to make refinements and changes in technology appropriate to their environment, but they were not the sort of innovations ABT wanted. This difference in outlook is clear in the case of Charles McCluer, one of the more inventive minds at the southern company.

A superintendent of the Richmond exchange, McCluer had been a key technician early on at SBT&T. Between 1889 and 1892 he produced and asked American Bell to patent a common return wire to reduce induction noise; a busy test for switchboards; a compound metallic circuit; and an automatic telephone toll box.[27] None of these innovations constituted a breakthrough.[28] Each was a small adaptation, a clever but not very sophisticated solution to a limited problem which had arisen in the course of McCluer's routine work in the South. His coin box telephone, he admitted himself, was most useful in "small towns and villages that cannot support regular exchanges."[29] It provided a simple means of connecting subscribers and was used mainly in sparsely populated regions of the South and Southwest where exchanges were not practical.[30] His busy test and common return wire, designed specifically for the Law system, were employed by Southern Bell only in exchanges using this equipment.[31]

Each of McCluer's innovations bore the mark of the southern economy, placing them in conflict with the plans for technological control and coordination in evolution at American Bell. The plans of the parent company called for more expensive equipment and greater capital investment to prepare the way for long-distance service. Seeking to expand its operations in profitable urban markets in the Northeast, parent company engineers devoted their time to the complex problems of high volume transmission and long-distance communications. They worked on new switchboards of greater capacity, underground conduits to relieve overburdened poles, and new types of circuits for protection against induction noise. Such matters were of far less concern in the sparsely settled South. As a result Southern Bell personnel like McCluer devoted their energies to other matters, developing an opposing technological style.

This basic difference of styles brought the case of Charles McCluer to a sad denouement. ABT refused to patent his innovations or compensate him for his work, as Bell patent expert Thomas Lockwood rejected each of his devices as too specific for further development.[32] While some members of the Bell inner circle wanted to buy and patent even small modifications—for they added to the total stock of technical knowledge—Lockwood was more selective.[33] Throughout his report, he disparaged the quality of McCluer's work and questioned the man's technical competence. But these were not the real issues. All of McCluer's inventions worked; they were just confined to the particular needs of regions like the South. He built inexpensive, easy to operate equipment which could be employed with only a small capital outlay. These qualities, Lockwood believed, ran counter to the larger needs of the telephone industry, and he feared that a more liberal policy toward McCluer would encourage further deviations from Bell standards.[34] McCluer, of course, did not see things this way. Growing alienated from his employers because of what he believed was unfair treatment, he was eventually fired in a dispute at the Richmond exchange. He ended his life an embittered old man who continued to harangue AT&T executives for years to hear his case and make amends for their predecessors' insensitivity.

As the case of Charles McCluer makes clear, a technological style is not simply a matter of discreet inventions, but how those inventions fit a whole system of technology. One new device may stimulate research in a new field of engineering. Similarly, solving even one problem may require redesigning an entire system of interlocking equipment. Northern and southern telephone men differed not only on their opinions of new telephone equipment, but in their method and approach to research and development. They defined different problems to be solved, different "reverse salients" to future growth, and as a result, developed their different technological styles.[35]

The managers and engineers of ABT understood quite well these implications. To raise long-distance transmission quality, for example, AT&T had to improve long lines themselves, and redesign local equipment and reroute local traffic patterns to fit this improved technology. It could not pursue this policy if operating companies continued to choose and develop technology independent of Bell's central directives. This perception lay behind Hammond Hayes's remark that one "mistake" in choice of equipment would lead to another, eventually threatening the whole design of the Bell system. To prevent such mistakes from occurring, parent company executives began to restructure relations between themselves and managers of their regional firms. In matters of technology in particular, ABT, and its increasingly important long-distance subsidiary AT&T, began to assert a much stronger hand in shaping the views of operating company personnel.[36]

Early on some members of the Bell Company had recognized that the different perceptions of managers in different regions might lead to a variety of technological styles in the telephone industry. Their original fear was that the shortness of the original five-year license the parent firm granted its operating companies would stifle innovation in the field.[37] AT&T vice president E. J. Hall believed that without such innovation, expansion would become impossible.[38] Many necessary refinements in technology were quite specific and had to be devised in practice, rather than designed beforehand. As Bell did not have the technical or financial resources to carry out such innovation itself, it depended on the talents of field technicians and managers. Hall proposed, therefore, that the company abandon the five-year contract for a longer one, and institute some method to monitor field innovations and diffuse generally applicable improvements throughout the Bell organization.[39]

The limitations of the five-year horizon were solved when Bell began to issue perpetual licenses and acquire an interest in its operating companies. Yet as Hall and others had perceived, innovative activity was affected by economic incentives, incentives that varied by region, as we have seen.[40] To overcome this larger problem, the parent organization took a greater role in disseminating information and coordinating innovative activity in an effort to enforce a uniform policy on technology.

The first procedures of this sort were quite informal, involving meetings between members of the parent and operating companies to exchange thoughts and information. The most famous of these early programs were the technical conferences held in the 1880s and 1890s. They included the National Telephone Exchange Managers Conferences, the Switchboard Conferences, and the Cable Conferences. Initially they were democratic in nature, with all present exchanging ideas on technology and operations. After 1885, however, the meetings began to change. ABT technicians played a much more dominant role, and set out to correct deviations in practices and procedures which appeared at the regional level. The nature of the meetings and the presence of well-trained Bell company engineers gave the parent firm a lever, which it used to push for standardization in equipment design and use. Advocating universal standards of efficiency for telephone technology, Bell personnel began to replace field engineers' pragmatic and particular notions with their own ideas of standardization and systematization. These were the values of the parent company, however, and they allowed little room for consideration of local economic conditions.

Attuned strictly to the needs of their own economy, Southern Bell engineers could not produce innovations that met these requirements. As a result, they contributed little to the conferences. The majority of papers were given by Bell men from New

York, Chicago, and the Pacific Coast companies. At the Seventh Annual Exchange Managers Meeting, Charles McCluer did present a paper on ways of reducing induction noise from railway lines.[41] This was an important problem, but McCluer addressed an aspect of it relevant only to the Law system, which was in use at his Richmond exchange. By focusing on a problem of such narrow scope, he had stepped outside of the conference themes of standardization and best practice.[42]

By 1889, these themes were becoming dominant throughout the Bell organization. At the Eleventh Annual Exchange Managers Meeting, J. J. Carty, of New York's Metropolitan Telephone and Telegraph, proposed that Bell was entering a "new era of telephony." Carty outlined the implications of this new era for telephone technology. He maintained that "although individual discoveries and inventions will undoubtedly lead to improvements . . . it will be by adhering to uniform practices, by a concerted effort on the part of companies taking up the new service and careful evaluation of their experiences, that the greatest improvements may be looked for."[43] Carty then went on to deplore the lack of nationwide fixed standards in the industry and the persistence of significant variations in equipment between regions.

Reaction to this paper was startling, in the context of the normally sedate meetings. Some operating company managers praised the presentation and asked that it be published for all to read. Others, however, wanted it suppressed and felt that ABT should disclaim responsibility for its contents.[44] Thomas Lockwood, the Bell patent advisor, tried to appease both sides. But he ended up claiming that uniformity was desirable in all aspects of the telephone business. Despite his claim, many members of Bell regional companies were not ready to give up the pluralism which had characterized parent-operating company relations. To them, variations in conditions were still important and legitimate criteria for choosing equipment and designing technology.

The South expressed its reaction to the "new era" in 1892 at a conference of southern exchange managers and superintendents. Present at the meeting were ABT representatives Thomas Lockwood and William Thompson.[45] The parties assembled held distinctly different views of the future of southern telephony. Individual southern exchange managers were deeply concerned with conditions in their own locality. Executives of SBT&T itself took a broader regional view, but still focused on the South. Lockwood and Thompson, on the other hand, spoke as agents of ABT. Lockwood in particular was an expert on operating company technology and ABT's attempt to control it.

Although the meeting was billed as an exchange of information about the latest ideas in the business, another text ran beneath the surface. In this amiable forum the assembled parties addressed the changing relationship between parent and operating company. The new hierarchy, running from ABT to the local level, was already apparent. American Bell had assumed majority ownership of Southern Bell, as well as most other regional firms. Its subsidiary, AT&T, was building transregional long lines. Plans such as those of Carty were already in the works to tighten up the informal structure of Bell operations.[46] But actual control had still not passed to the parent firm. As the meeting would reveal, there was still room for debate as to which policies would be followed.

As with most such meetings, the days were filled with extended introductions, dry papers of interest only to insiders, and prolonged, congratulatory question and answer sessions. But when the participants stepped back a bit and surveyed the whole scene, they revealed something of the differences which divided North and South. Thomas Lockwood's long, general speech near the end of the meeting stressed the need for cohesiveness between ABT and its southern company. He chided Southern Bell managers for their misunderstanding of the industry, a misunderstanding which too often found its way to the public and state legislatures, fueling resentment against Bell. Lockwood himself stressed Bell's consistency, steadiness, and commitment to high-quality work. Implementing these policies required "reasonable" rates and standardized practices, especially in the choice and use of equipment. Engineer Thompson reiterated this last point when he enumerated rules for inspectors and technicians to follow to insure quality work.

Though his remarks concerned the mundane chores of day-to-day telephone service, Lockwood hinted at broader themes. He concluded his speech by linking standardized local service to the growth of long-distance telephony, and to improvements AT&T

was making in induction coils and long-distance transmitters. He also noted the difficult time the long-distance company was having in getting regional firms to move with vigor into the long-distance business.[47] Behind these issues was the new parent company attitude toward its operating units, Southern Bell in particular. Lockwood revealed this attitude in his somewhat patronizing tone and his belief in his own superior understanding of the technical needs of the business. Filling his speech with homey examples and metaphors, he made his point clear: The southern firm needed ABT's guidance to avoid errors and erase technical, managerial, and political mistakes. High-quality work and consistency of performance were the goals, but the standards for both were to be set by ABT. These standards would be expensive to achieve and would require "reasonable" rates. But mostly they would require operating companies to follow the dictates of the parent firm.

That the southern managers might have had different concerns, set different goals, or held different standards seems not to have concerned Lockwood. He did not have to face the cost of employing a new technology or the difficulty of maintaining and servicing it. He did not have to deal with customers in small towns who wanted but could not afford telephone service on these terms. He did not have to worry about investing in equipment compatible with long-distance telephony in a region which had very little demand for such service.

The members of Southern Bell had their chance to respond to Lockwood's remarks, and they did so without hesitation. Lockwood's proposals for ensuring quality and consistency meant eliminating local variations in technological and managerial decisions. But in a realistic appraisal of conditions as they existed in the South in 1892, SBT&T secretary Daniel Carson noted, "every exchange manager is in business for himself."[48] Though he was stressing the importance of managers taking strict account of profitability in their work, Carson's statements reveal just how the southern firm—and many other regional Bell companies—was organized. Its units, exchanges, were basically independent enterprises. They were joined to Southern Bell legally and financially, through the firm's exclusive franchise with American Bell and its control of capital. But day-to-day decisions were left up to exchange managers, the backbone of the company. They decided what local conditions required and acted accordingly. Southern Bell employed only a small central staff to coordinate their activities and set general policy.

James Easterlin picked up Carson's theme in a paean to the exchange manager. His typical southern exchange was located in a small town; its manager was a local boy, well known to the town's commercial, landowning, and political elite. Intent on serving their needs, the manager did not reach beyond his means. Easterlin satirized the "puffy" exchange manager whose ambitions went beyond the requirements of his local clientele. Such a man, seeing growth and change everywhere, made imprudent increases in expenditures, much to the detriment of Southern Bell. W. T. Gentry provided the final word on the matter in a speech on the telephone business of his home state of Georgia. Gentry noted the southerness of Southern Bell. Despite its incorporation in New York, its close connection to ABT, and its outside investors, the firm was, Gentry maintained, at heart a southern enterprise.

Through their meeting, the executives of Southern Bell and representatives of ABT had defined two opposing views of the telephone business. Gentry, Easterlin, and Carson stressed the southern nature of the telephone industry in the South, its responsiveness to local conditions, and its service to the regional economy. Lockwood cautiously acknowledged the value of loyal, southern-born managers, but, choosing his words carefully, tried to instill in them a sense of loyalty to ABT and its grander designs. Where Easterlin stressed the need for a modest and cost-conscious approach, Lockwood called for greater expenditure and a commitment to the superregional needs of the telephone system. Where Gentry praised the southerness of Bell in the South, Lockwood proposed tighter organizational control from the parent company. Where Carson emphasized the independence of the exchange manager, Lockwood outlined a hierarchical structure to eliminate variations in equipment choice and business policy. During the next ten years, these opposing views would come into increasing conflict.

Resistance by Southern Bell managers did not slow the parent firm as it pushed ahead in the early 1890s with its efforts to control regional company technology. By then the company had another motive as well. It was under pressure from the approaching expiration of its patent on the telephone. Bell sought to both protect its market from new entrants and to expand long-distance service with its new policies on technology. Despite wide differences of opinion within the Bell organization, ABT began to reorganize its engineering and research departments so that they could exert greater influence over operating companies. These departments began to combine their earlier work in patent protection with enforcement of standards of design and practice.[49]

Going beyond the reforms of the 1880s, ABT's engineering department began in the 1890s to design many local exchange facilities. At the end of the decade, ABT and AT&T established central research and engineering departments. At AT&T, engineer Joseph Davis worked to overcome the problems of nonstandard equipment in the field in order to increase the efficiency of the long-distance network.[50] At ABT Hammond Hayes proposed that a committee of Bell and Western Electric engineers be formed to review the plans of operating companies.[51] Neither Davis nor Hayes was much concerned with theoretical research.[52] Rather, they wanted to prevent the subversion of Bell's plans from within by wayward operating companies. Therefore, they proposed changes that made relations between parent and operating company more hierarchical.[53] ABT centralized important decisions regarding technology, investment, and expenditures, which it believed could not be safely implemented at the local level if the long-distance network was to grow.[54]

Policies such as these altered some of the supply-side conditions that had led Southern Bell to its path of technological development. Though the top management of Southern Bell came from outside the region, the company drew its lower level managers and technicians from within the South. People with the necessary technical skills for these positions were in short supply in the region. This shortage northern observers tended to attribute to the South's lack of "native genius," but the real culprit was the region's comparatively poor quality education, its overwhelming rural poverty, and its small number of industrial firms. Without the heavy industry and profitable markets of the North, the South could neither attract nor generate skilled telephone workers. This lack of trained engineers seems to have been especially acute compared to other regions. In contrast to the rural Midwest, the rural South lacked nearby cities and industries, and had made little improvement in its system of technical education since the Civil War.[55] For the most part, southern industry remained small in scale, simple in structure, and local in character. All of these conditions contributed to the South's inability to adopt complex new technologies.

While greater top-down control by ABT and coordination of regional company technology by Western Electric eliminated some of these problems, these reforms failed to address Southern Bell managers' legitimate market concerns. Southern customers had barely gotten used to telephones themselves, and it was far from clear in the 1880s they would have any use for or be willing to fund a long-distance network.[56] The region's dispersed rural population and small urban sector discouraged the use of equipment designed for the dense urban markets of the North. Yet the parent Bell organization, secure in its monopoly, refused to compromise its technological vision.

The difficulty of integrating the South into the evolving telephone network went far deeper than most parent company personnel appreciated. Bell managers tended to blame southerners themselves for failing to appreciate the advantages of the system approach and for squandering opportunities to market the telephone along the same lines as was being done in the Northeast. There were clear differences between the southern and northern business styles; as we have seen, in the early years the South needed assistance to avoid pitfalls and market the new technology. As became clear by the late 1880s, however, more was involved than the entrepreneurial shortcomings of southern businessmen. The economy of the South, overwhelmingly rural and lacking the lines of commerce, transportation, and communication that bound together places in more advanced regions, set limits on how far the system could be extended in the South in this period. ABT managers would soon discover that top-down control and coordination could not eliminate these demand-side realities. Competition ultimately drove home this point, but even before the birth of nationwide competition in 1894, Bell had a foretaste of the profound difficulties it would face in erecting its system in the South and similar regions.

In 1883 J. Harris Rogers, engineer for the United States Capitol, invented a new telephone transmitter which quickly found favor with some members of the southern business community.[57] Since the formation of Southern Bell, indigenous southern entrepreneurs had been shut out of the growing telephone industry. The only way they could enter was by circumventing Bell's patents. In Rogers's invention, some believed that they had found a means of doing so. By 1887, SBT&T's annual report conceded that a few scattered "infringers" using the Rogers technology were draining some revenue from the company's private line business and had set up exchanges in several towns.[58] By 1889, with backing from several United States congressmen, a new firm, the Pan Electric Company, had been founded. Using Rogers's invention, this company began to construct long-distance lines in the Washington, D.C., vicinity. A little while later, Charles Long, a founding member of Pan Electric, broke away to start yet another competing company. Employing an alternative to the Bell telephone he had invented, Long set up operations in selected spots in the South as well.[59]

The new entrants did not greatly reduce Southern Bell's revenues. Neither firm had a viable alternative to all Bell equipment, and neither had succeeded in raising a large amount of capital. But their efforts sent shock waves north to American Bell, which moved decisively to quash the challenge. Bell responded quickly because it saw the new firms as a threat to its basic telephone patents, which were primarily responsible for the firm's monopoly position.[60] More generally, it recognized that the new technology was also a threat to its vision of an interconnected system. In 1888 ABT began infringement suits against the Long and Rogers devices. Eventually decided in Bell's favor, these suits destroyed any hope the new firms had of competing.

As this episode reveals, the Bell Company was willing to take aggressive action to defend its technological style. Bell might have put an end to the challenge quietly, by acquiring the patents of the new companies. Their competitive claims were small, and their efforts bordered on fraud. Since 1880 the company had bought many patents and innovations, to protect its market position, create new services and sources of revenue, and prepare the way for the growth of long-distance communications. It had little interest in acquiring the Long or Rogers technology, however.[61] Beyond stifling future competition, the value of a patent acquisition was measured by its contribution to Bell's storehouse of useful technical knowledge. Neither competing device met this requirement; rather, they constituted challenges to the technological system Bell was assembling. Purchasing such equipment would have only encouraged further deviations from the company's policies.

The belief that the competitors were a threat to Bell's technological style was no mistake. Both of the rivals sought to market their technology as an alternative to Bell's. The Long firm operated in marginal areas, avoiding direct confrontations with Bell in the South. It found untapped pockets of demand in North Carolina, Virginia, Kentucky, and Alabama, as well as Louisiana, Arkansas, and Texas, where it brought cheap, rudimentary service to places deemed not profitable by Southern Bell. Eventually both of the firms would have had to join the Bell system to survive. Pan Electric's backers admitted that they entertained hopes of establishing contractual relations with ABT and serving as another source of equipment to Bell licensees.[62] But by offering a different type of equipment with appeal to a market Bell did not yet serve, they threatened to undermine Bell efforts to standardize telephone equipment nationwide.[63]

Coming in the South, this challenge was especially worrisome. Bell's control of that region was still somewhat weak. Even in 1890, after it had acquired majority ownership of Southern Bell, Western Union remained a powerful minority stockholder, and it was still involved with Bell in a dispute over the terms of the 1880 settlement. Southern Bell itself had already experienced a "pretty row" in Atlanta with the Browns and was wary of independent-minded local entrepreneurs. New firms offering alternatives to Bell technology were too dangerous to take lightly.

The new entrants also frightened Bell by using legal ploys and political pressure to break into the telephone business. Bell was sensitive to the charge of monopoly, and the Long and Pan Electric firms played on this fear by calling themselves "home" institutions fighting the "grinding" Bell monopoly.[64] Proffering cheap alternatives to Bell service in marginal areas of the South, Long found consumers who objected to Bell as an expensive, yankee-owned enterprise. Part of the Pan Electric backers' plan was to combine public sympathy with legal maneuvers to delay Bell's response long enough to establish a foothold in the South.[65] The firm bolstered its position by making President Cleveland's attorney general, Augustus H. Garland, former senator from Arkansas, a major stockholder. He obligingly filed suit to annul Bell's patents.[66]

Though this obviously corrupt effort was soon snuffed out, the Pan Electric and Long firms continued to rely on powerful political friends and support from dissatisfied Bell customers to make headway. The backers of these enterprises had rather modest goals—either to capture a small slice of the telephone market, force Bell to purchase their claims, or profit by stock speculation in their ephemeral enterprises. Bell viewed these practices as blatantly fraudulent, though they were actually rather typical of Gilded Age schemes. Moreover, while Bell characterized the Long and Rogers devices as deviations from standard telephone technology, southerners saw them otherwise. To southern businessmen who had been shut out of the industry, they were an opportunity; to southern consumers who could not afford Bell service, they were an inexpensive alternative. The equipment was particularly attractive at the margins of the telephone market, where the cost of Bell service exceeded the means of most customers.[67]

It was the combination of public sympathy and political support that evoked Bell's swift reaction. Ignoring the threat would have, in the eyes of Bell managers, only led to the proliferation of cheap equipment attractive in poorer regions like the South, slowing the advance of Bell telephones and undermining AT&T's plans for construction of a national telephone system. Against such possibilities Bell chose to flex its muscles and crush the rebellion immediately. Taking the high ground, the company resorted to the courts, refused offers of a compromise, and put a stop to the challengers.

This early competition was only a minor note of dissent in what was a thoroughly monopolized telephone industry. Because of Bell's control of patents on the telephone and important central office equipment, no outsider could gain entry for long. But the incident foreshadowed later problems that would arise in the South with the end of Bell's basic telephone patent. Wide variations in regional market conditions made the creation of a national network as Bell managers envisaged it a difficult prospect. Although standardization, integration, and homogenization of regional communications service seem from our perspective perfectly natural, even inevitable, these ideas were radically new in the late nineteenth century. No one in 1885 could prove that an integrated system was the best way to proceed. Bell's commitment to this system reflects the technological mindset of its key managers and engineers.[68] Pursuit of this technological ideal in the face of adverse economic conditions proved to be one of the major quests of the company in the early decades of its history. Beginning in 1894, the road to success became considerably more difficult as the firm's patent wall crumbled.

James Merrill Ormes, the entrepreneur who founded Southern Bell. Ormes was the key innovator in the early southern telephone industry. Photograph courtesy of Southern Bell Telephone Company.

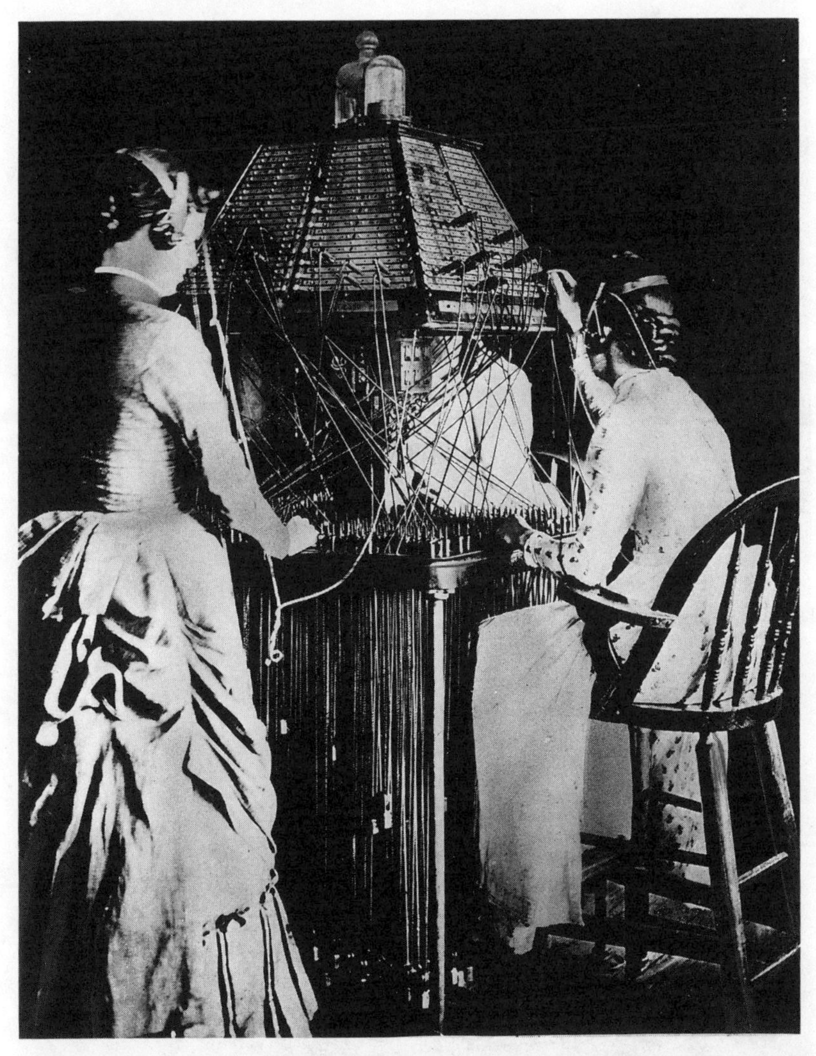

The Law telephone switchboard in use in Richmond, Virginia, 1882. Though early on it was favored by southern exchanges for its speed, low cost, and ease of use, the Law board ran up against strong opposition from American Bell.

The Law board in St. Louis, Missouri, 1888.

Early Southern Bell central office in Atlanta, Georgia. In the beginning, Southern Bell had to rely on what was available; later it would construct its own buildings following specifications laid down by the parent Bell organization.

Charles Jasper Glidden, 1900. At one time his Lowell Company controlled considerable telephone property in New England and helped to bring about the consolidation of the industry in that region.

Thomas D. Lockwood. Though not a lawyer, Lockwood was Bell's expert on patents, and he had very definite ideas about the control and use of new technology in the Bell system.

CHAPTER 5

Market Challenges

BUILDING A NATIONAL, interconnected system in a competitive market was the greatest challenge the Bell Company faced in the early twentieth century. In regions like the South in particular, competing firms responding to market conditions began to promote other, more restricted types of telephone service at variance with Bell system designs. During the monopoly period, Bell had been able to turn aside these alternatives. The company had to modify its strategy after 1894, however, when the expiration of the Bell telephone patents allowed the entry of numerous new competitors—the so-called independents—into the industry.

These important policy revisions took place slowly, as Bell managers at first did not understand the nature of the challenge they faced. During the years of monopoly, the company had kept prices high and accepted slow but steady increases in rentals.[1] Concentrating on dense, profitable markets, where its technology worked best and from which it derived the bulk of its income, the corporation had invested heavily in sophisticated equipment that paved the way for long-distance communications.[2] Though this strategy enabled Bell to successfully market telephones in cities, it prevented the company from penetrating more remote places at the margins of urban America. High prices and restricted service left untapped pockets of demand in rural and small town markets, which independents began to serve. In such places they provided a variety of types of service and equipment. Undercutting Bell prices, the competitors secured a substantial share of the telephone market by 1900, slowing expansion of the Bell system in the South and elsewhere.

The competitors were able to undersell Bell for two reasons. As a monopolist, Bell had restricted output, leading to higher prices. The entry of the new firms very quickly put an end to that situation. The new firms also succeeded, however, because they responded to Bell's high level of product quality, which had also contributed to its high prices. The company's commitment to its system led it to use premium equipment and undertake complicated feats of engineering to extend that system and keep all parts of it in continual operation. The more extensive the system, the more the issue of quality mattered. By offering a wider range of equipment and service, some less expensive and lower in quality than Bell's, independent firms gained customers unwilling to pay for Bell's superior quality service.

Bell contributed to the success of its competitors by oversupplying the optimal level of quality to the market during the monopoly era. Like most firms, the corporation planned the quality of its product on the basis of prices and output (quantity), setting product quality equal to the preference of the marginal consumer.[3] For economic efficiency, however, quality should match the preference of the average consumer. A firm, particularly a monopolistic one, can easily miss the welfare maximizing level of quality.

To determine if a company oversupplies (or undersupplies) quality, we must know the values of two parameters: The relationship between quality and quantity; and, in the case of a monopolist, the degree of monopoly output restriction. The first value may be either positive or negative. If the marginal value of quality to consumers rises with quantity, the relationship is positive. If it falls, it is negative. A positive relationship implies that each additional consumer values quality more than the preceding ones. A negative relationship implies the opposite—that each new consumer values quality less than those who preceded him. The second parameter—the degree of monopoly output restriction—may be either large or small. A restrictive monopolist holds output far back from the competitive equilibrium. A nonrestrictive monopolist produces near the competitive equilibrium. The interaction of these two parameters determines the level of quality a monopolistic firm will supply.

A restrictive monopolist operating in a market with a positive quality-quantity relationship, for example, will undersupply quality; in a market with a negative relationship, the same firm will oversupply it. Conversely, a nonrestrictive monopolist will oversupply quality if the relationship is positive and undersupply it if the relationship is negative.

These parameters are difficult to evaluate for the telephone industry. Telephone rentals increased dramatically after 1894, suggesting that Bell had been a restrictive monopolist. Increases were lowest in New England and greatest in the South and Midwest, where the new competitors were concentrated. It seems likely then that Bell was a restrictive monopolist in these latter markets, whereas its prices and quality conformed fairly closely to what New England consumers desired. The quality-quantity relationship in telephony is even more difficult to determine. Today, there is reason to suspect that demand for quality and quantity increase together—that the relationship is positive. An increase in the number of subscribers raises the utility of the telephone and stimulates demand for better service, more connections, greater range of communications. These changes in turn imply an increase in telephone quality through better transmitters, more circuits, and sophisticated system engineering.

Bell managers in the 1890s seem to have proceeded on the assumption that demand for quality would rise with the expansion of service.[4] They had a very precise idea of what telephone users wanted. In their view, good service was that with which "non-experts could converse about subjects with which they were not familiar without repetition."[5] Lying behind this definition was the image of a world of businessmen, engineers, and professionals communicating about technical matters with peers with whom they were not intimately acquainted. Such a world demanded a high-quality national long-distance system because its residents had many and distant correspondents and contacts.[6] This world, the one that Bell managers were trying to serve, existed primarily in large cities in 1894. It was not that of the majority of potential telephone customers.

The picture that Bell engineers painted of quality service spoke very little to the needs of farmers and small town residents in the South and elsewhere. Many, perhaps most Americans in the nineteenth century had communications needs that were almost strictly local. The information they exchanged concerned specific community matters such as the weather, the placement of orders at stores, or news about families and friends. Responding to Bell efforts to improve service and extend long-distance service in Huntsville, Alabama, for example, a local editor wrote that lower rates and local connections to nearby villages and farms were what citizens wanted, not lines to New York or Washington.[7]

As Bell managers would only slowly come to realize, these communications needs required considerably less sophistication in equipment and line quality than Bell was preparing to provide. Only after difficult years of competition did the company see that in rural and semiurban areas, comprising 65 percent of the nation, demand for service differed considerably from the urban areas Bell primarily served.[8] Widely available local service, even if limited in range and "poor" in quality by Bell's definition, had the advantage of being low in price, something of prime importance to consumers in such places. Indeed, as the man who had formulated Bell's definition of good quality service himself admitted, "just what constitutes poor service is a question upon which we may not all agree."[9]

If, as seems likely, in many places new customers' first concern was inexpensive local service, not integrated local and long-distance service at premium prices, then quality and quantity demand did not rise together and Bell had oversupplied the optimal level of telephone service quality during the monopoly period. Under these conditions, competition became a battle over the nature of telephone service, with Bell offering high-quality service at premium prices, and independents offering lower-quality service at lower prices.

Clearly many forces contributed to the success or failure of new telephone firms. Local resentment against the eastern telephone monopoly, assistance from local politicians and business elites, and simply the vagaries of a newly competitive market aided telephone independents. Just as surely, limited funds, entrepreneurial conservatism, and inexperience hurt the small-scale competitor. Overall, however, it seems that the fate of new entrants hinged mainly on the question of quality. Customer preferences for quality of service set the dividing line between the Bell and independent territories. Where customers opted for lower-priced, limited service, the new firms became thoroughly entrenched. Where Bell was able to convince customers to accept its brand of service, it retained market control.

The patterns of competition in the telephone industry support this case. The majority of new firms were small, often local concerns bent on underselling Bell in their home-town market.[10] Although some competitors arose to challenge Bell at its strong points—the nation's large cities—few succeeded unless they could exploit a weakness in Bell structure or gain political support from city and state governments anxious to discipline the Bell monopoly.[11] In most of the economically advanced Northeast and Pacific Coast, competitors were confined to the rural and outlying counties. In heavily urban Massachusetts, Connecticut, and Rhode Island, Bell controlled even hinterland areas (see table 5.1). In these states, customers apparently valued Bell service and were willing to pay premium prices for it. With its long head start in meeting these demands, Bell had a competitive edge over its rivals. Its many intercity lines, combined with the efficiency of its equipment in large urban markets, made it difficult for new firms to compete.

Independents proved successful when they operated in places with different demand conditions, in areas that had a lower per capita income, a larger rural population, and less industrial development than those dominated by Bell.[12] In such places, competitors were able to take advantage of Bell's rigid price structure and higher than needed service quality to carve out large markets for themselves. They offered lower quality, rudimentary service for certain classes of customers Bell had ignored. Because independents did not have to construct an interconnected system and because they had much to gain through price competition, they were able to undersell Bell.

For these reasons, the South became one of the focal points of competition. One hundred fifty-five independent firms served the South. Supported by small town boosters and regional promoters, they cut drastically into Bell's market share, reducing it to less than half by the early twentieth century. By 1902, the states in which Southern Bell operated, excluding West Virginia, had 7.2 percent

Table 5.1. Bell and Independent Telephone Customers, 1902

	South			New England			Midwest	
	Ind.	Bell		Ind.	Bell		Ind.	Bell
AL	3,310	8,767	CT	774	21,720	MI	44,119	49,842
FL	5,255	2,961	MA	2,911	93,601	IL	95,118	116,069
GA	11,521	14,169	ME	1,766	12,279	IN	95,120	37,369
NC	13,367	2,685	NH	1,555	8,394	IO	96,728	23,269
SC	6,921	3,546	VT	5,675	6,439	OH	128,784	93,983
VA	15,170	8,960	RI	166	10,708			
WV	16,501	5,875						
	72,045	46,963		12,681	153,141		459,889	320,532

Distribution: Percentage of Total Telephones in Each Region

	Independent	Bell
South	7.2	3.5
New England	1.2	11.6
Midwest	46	24
Far West	6.2	5.4
Pacific	0.5	11.6
Mid Atlantic	14	38
Total	75	94

Source: AT&T box 1116, William Allen, History of the Independent Companies.

Note: Total Customers: Independent, 998,119; Bell, 1,317,178

of all independent telephones, compared to only 3.5 percent of Bell telephones (see table 5.1). In comparison with other regions, the southern response to competition was even more striking. Only forty-two independent firms operated in New England, less than 2 percent of the nationwide total. The New England competitors supplied only 1.2 percent of the non-Bell telephones in the nation, moreover, while Bell had 12 percent of its instruments in the region. In terms of total numbers of telephones, independents were strongest by far in the Midwest. By 1902, Bell competitors in Michigan, Illinois, Indiana, Iowa, and Ohio had managed to rent or sell 439,889 telephones to Bell's 320,532, exceeding Bell by some 69 percent. While the raw figures are less impressive for the South, there independent telephones exceeded Bell's by a similar percentage, 65 (see table 5.1).

If we examine the subsectors of the southern market, we can see that the issue of quality was the most important determinant of independent firm success. Competing telephone companies made some inroads into the more industrial and urban sectors of Alabama, Georgia, and Virginia, where incomes were relatively high and demand for telephone service was generally strong. But the competitors did not last long in these areas, because demand for Bell's more extensive service was even stronger. Local firms may have provided a sense of comfort and insularity for beleaguered "island communities," but they faced a stiff challenge from Bell when their customers began to demand outside connections. By 1902, Bell telephones in Alabama and Georgia exceeded those of independents. In Virginia, Bell was still behind, but was far closer to parity than in other southern states (see table 5.1).

Using its competitive edge in the more industrially developed southern states, Bell achieved some important early victories over its rivals. In Selma, for example, local capitalists tried to compete with Bell, but the company was able to defeat them soundly. E. D. Bowles, cashier of a Selma bank, formed a company to market the Harrison telephone, a Bell alternative, in his city. He received managerial support from bank president E. M. Wilson and backing from three other prominent businessmen in the community. In 1896 they formed the Alabama Telephone and Construction Company.[13] The new firm opened an exchange in Selma, but it attracted only 132 customers. The company also moved into Atlanta, but enjoyed little favor in this important New South metropolis either. While a nuisance to Southern Bell, the firm's initially poor showing prevented it from attracting the capital needed to mount a serious challenge. As one backer, George Wilkins, admitted, the low rates the firm had to charge to enter the market could not sustain profits. "It is all well and good to talk about home institutions and local influence," he wrote, "but the investors want and expect to receive some return on their investment."[14] Bell was able to quickly put an end to the Alabama venture through direct competition, destroying the opposition telephones "after preserving the most useful parts."[15]

Throughout the relatively advanced sectors of the South, Bell took advantage of demands for its high-quality service to outcompete independents in this way. In Mobile, for example, the firm laid expensive underground cables and added metallic circuits to its exchange. Town officials and telephone customers responded favorably to this work, placing pressure on independent promoter Adam Glass to answer in kind. Unable to follow Bell in upgrading service, Glass was forced to sell out.[16] New entrants did little better in Birmingham, another important New South city.[17] An independent telephone company in the Alabama steel center managed to attract 826 customers by 1900, and added over 400 more a year later.[18] This early success obviated a quick Bell victory, preventing the company from purchasing its rival for fear of encouraging more competitors. Nonetheless, Southern Bell was able to defeat the challenger by applying pressure indirectly, through its long-distance connections. Bell enlarged its switchboard capacity in Birmingham, providing more space for long-distance trunks, and used expensive metallic circuits to improve transmission quality. By upgrading the quality of its local service and building lines to surrounding communities, Southern Bell quickly isolated the Birmingham independent. With its superior regional toll network, the company prevented the competitor from building its own intercity lines and joining forces with independents in Mobile, Bessemer, and Tuscaloosa. Bowing to Bell pressure, the Birmingham rival left the business with only a modest profit.[19]

In a similar manner, Bell lured smaller towns in the area such as Tuscaloosa and Bessemer into its fold with promises of connections to Birmingham. Because the area bounded by Birmingham, Pensacola, Atlanta, and Montgomery had a substantial amount of commerce, a relatively high population density, and some heavy industry, Bell was able to use its up-to-date technology to construct long lines and form an interconnected system throughout the region. This effort proved effective against independents, though the firm had a close race with a rival, the Southern States Telephone Company.[20] Using capital from the Viaduct Manufacturing Company of Baltimore, this independent tried to compete by organizing local telephone firms into a regional system.[21] But Bell's superior resources and knowledge of long-distance telephony gave it the edge it needed to win. Other telephone promoters in Georgia and Alabama also tried to build their own long lines to meet this challenge, but Bell's dominance of the intercity market proved too strong an advantage for them as well. With few exceptions, independents were unable to overcome Bell's lead in this type of service. Though as late as 1907 non-Bell firms in Georgia and Alabama controlled 40 percent of the market, Bell policy had severely curtailed their potential to expand and had reduced them to the marginal sectors of the states.

Although Bell managers hoped that toll lines could be used as competitive weapons against rivals in other parts of the South, they soon discovered the limitations of this strategy. Just north of the firm's stronghold near Birmingham, for example, was a wide open territory where demand for Bell's type of service was weak.[22] Here the company could not achieve dominance, despite its well-developed nearby toll network. In those places where Bell had no such network, conditions were even worse. In North Carolina, South Carolina, and Florida, as well as rural subsectors scattered throughout the seaboard South, local companies effectively outcompeted Bell.

Bell's most difficult challenge came from the well-organized local firms in North Carolina.[23] Like the area between Atlanta and Birmingham, the Piedmont district of North Carolina grew rapidly between 1880 and 1900.[24] Prosperity characterized its midsized market towns, upcountry resorts, and rich agricultural areas. In Durham, J. B. Duke was organizing a major tobacco oligopoly. In Greensboro and Charlotte, progressive business interests were busy promoting town growth. Throughout the area, merchants and businessmen willingly embraced new industry and invested their surplus funds in money-making opportunities. After 1894, they quite naturally turned to the telephone. By 1902, independent telephones in the state outnumbered Bell telephones five to one, the highest ratio in the South, and probably the nation (see table 5.1).

As in other cases, this competition rested as much on demand as on supply-side factors. One of the most successful North Carolina independent enterprises, for example, the Interstate Telephone Company of Durham, was founded by local businessmen dissatisfied with high-priced Bell service. The firm quickly expanded and by 1900 was operating exchanges in Durham, Raleigh, Goldsboro, Winston-Salem, and Charlottesville, Virginia, as well as smaller points in between. The competitor neglected neither local nor long-distance demand. Its interexchange lines stretched as far east as Wilmington, as far south as Elkin, and as far west as Warrenton.[25] The company earned a gross income of $36,671 in exchange and $1,500 in toll receipts in 1900, generating a net income of $21,508. Based on the rate of subscriber growth, Southern Bell managers estimated that its net income would rise to $32,000 by 1901.[26]

The Durham firm succeeded for two reasons. First, it kept local rates low, something which appealed to the frugal customers of small towns.[27] Secondly, it used toll lines to link both its own exchanges and those of other, strictly local independents in towns such as Wilson and Goldsboro. These municipal companies, started as "home" enterprises in protest of high rates and "foreign" control would have otherwise remained isolated. Lacking access to outside connections, they would have become easy targets for a Bell takeover. The Durham telephone promoters, who eventually gained support from James B. Duke, were able to resist Bell by creating their own regional system.

Unlike Bell, Interstate did not construct its regional system with a national network in mind. Instead it served small towns which were, at this point, content with a few outside connections. Customers in such places were unwilling to pay for Bell's more fully integrated service.[28] These consumer preferences bring the issue of quality into sharper focus. The difference between Bell's service quality and prices and that of its competitors was not only a matter of equipment, but also extent of interconnected service. Bell's national network approach was generally more expensive than independents' regional systems. Though it kept its rates low, Interstate did not stint on equipment or capital outlays. Bell agents admitted that "the poles, phones, and switchboards are new and equal to those of any other system in the U.S."[29] But the North Carolina firm connected many points "of very little consequence and towns so small that it would not pay [Bell] to maintain an exchange therein."[30] By embracing the system approach, the company provided an effective alternative to Bell's long-distance service. By keeping its lines short, its network regional, and its prices low, it was able to force Bell into a compromise agreement in 1903.

North Carolina Interstate may have been something of an exception to the general rule of competition in the South. It served a market just right for its type of service, one neither so large that Bell could maneuver effectively, nor so small that there were no opportunities for profits. It also received backing from J. B. Duke, one of the few men in the South wealthy by national standards. But other North Carolina independents also prospered by finding submarkets beyond Bell's reach. The state, which had only seven Bell exchanges in 1902, supported 118 independent "central office systems," many exceedingly small.[31] Non-Bell firms provided profitable service in Newton and Shelby, for example, towns in which Bell would have lost money.[32] Conditions did not improve for Bell in the state over time. As late as 1911, SBT&T managers admitted that in North Carolina, "the opposition situation is perhaps worse, from a strategic standpoint, than any other state in which we operate."[33] As this report made clear, the competitors thrived by operating in places on the margin between high-quality Bell service and lower-quality local service. One group of independents, for example, clustered around the state's small manufacturing district near Taylorsville, Statesville, and Concord, where they were "very energetic" in toll line construction, much like the Durham company. Promoters in these places mobilized local capital, played on customer resentment of foreign corporations, and allied with small-scale manufacturers who had modest but steady demands for limited intercity service. Exploiting this combination of circumstances, they kept control of their market.

In other agri-commercial areas of the South as well, independents successfully followed strategies similar to those of the North Carolina firms. Florida too had an active agricultural sector which supported commercial enterprises, town growth, and some small-scale manufacturing.[34] In northern Florida Bell maintained market control, for this area was linked to the firm's successful long-distance network in Alabama and Georgia. In the "frontier" areas of south Florida and the Keys, however, local businessmen successfully promoted telephone service.

Businessmen and municipal boosters in West Palm Beach

and Miami, for example, joined forces in 1905 to construct their own local system. In the Palm Beach area, Dr. J. E. Liddy, J. R. Anthony, George Branning, J. B. Beach, and M. E. Gruber borrowed $2,500 from a St. Augustine bank and, with a city franchise, built their own telephone exchange.[35] Purchasing a second-hand switchboard, they placed it in a local drugstore and installed telephones in a nearby freight office. From these humble origins, their enterprise grew, as they strung lines to a local hotel and the residences or businesses of forty-two subscribers by the end of the first year. In the following years they upgraded their equipment, using an advanced common battery system.[36] In 1906, the Florida telephone pioneers contacted Miami businessmen, who wanted to link the telephone systems of the two towns. The Palm Beach independent continued to operate in south Florida until 1920, when Southern Bell at last purchased its exchange of 892 subscribers.[37]

In the rural South as well competitors exploited limitations of Bell service and technology. Profits were low in rural telephony, and Bell had all but ignored farmers, who, for the most part, were located beyond the edge of the Bell system and thus were expensive to serve.[38] To meet farmers' demands for less expensive service, Bell added some new wrinkles to its technology, such as cheap eight- and ten-party circuits. But this service was barely profitable for operating companies. It was also a nuisance. When hooked into an exchange, multiparty lines threatened to imbalance circuits and interfere with long-distance communications, thereby disrupting Bell's plans for integrated, high-quality service.[39] Yet to the rural residents, such service was highly valued. Cheap multiparty lines between farms provided rapid communications among neighbors over a radius of several miles; rural branch lines to town exchanges relieved farmer isolation.

The telephone, as it had been developed by Bell, simply could not meet these rural needs. An efficient means of urban communications, it was ill-suited to the demands of isolated agrarian life. A useful device for the daily exchange of business information among closely packed city dwellers, it was only marginally valuable to farmers who came to town once a week. Telecommunications assisted in the specialization of work routines in business and industry, but it offered far less to multifunctional farmers who operated their enterprises with little outside assistance. Even basic innovations such as the exchange could not be used profitably in sparsely settled agricultural areas. Newer equipment such as underground cables and metallic circuits, excellent responses to problems arising from urban congestion and long-distance transmission, offered little to farmers either.[40] Eventually some of the distinctive agrarian needs would be met by new technologies such as the radio and automobile. Others would be eliminated by the transformation of agriculture into argibusiness and the drawing together of town and countryside. In the meantime, however, farmers demanded a solution.[41]

As Bell technology did not permit one, farmers turned to independent firms and took advantage of the newly competitive telephone market.[42] Though southern farmers were the poorest in the nation, independent telephone promoters still found pockets of rural telephone demand scattered throughout the region. Responding to farmers' need for inexpensive service, they constructed simple systems.[43] Commercial independent firms kept costs low by providing farmers with telephones alone, and allowing them to erect their own poles and lines. Through such efforts some 4 percent of all commercial telephones (Bell and independent) were serving rural areas by 1902. Because of the way the census takers arrived at their figures, this percentage can be taken as a rough measure of rural residents served by small, non-Bell firms.[44] The enumerators considered rural all telephones of any commercial company whose principle exchange was in a city of less than 4,000 people.[45] Some farmers and rural residents may have been served by Southern Bell branch lines or by larger independents whose "principle exchange" was in a city larger than 4,000 people. In 1902, however, Southern Bell reported only 1,200 branch lines; and not all of them were farmers' lines. Some were private lines to outlying businesses.[46]

Even more successful in the rural South than local commercial companies were mutual telephone associations and farmers' independent lines. Run by farmers themselves, these organizations supplied rural residents with the cheap, rudimentary service they demanded. Farmers constructed multiparty lines among their neighbors and, where feasible, looped into a nearby exchange. Co-

Table 5.2. All Telephones, Type of System, 1902

	% Ind.	% Bell	% Mutual	% Farmer
North	24	75	1	0.5
Midwest	53	37	6.6	4
South	54	40	3	3

Source: Tables 5.1, 5.4. Bureau of the Census, *Telephones and Telegraphs, 1902* (Washington, D.C., 1906), table 38.

Note: North = MA, NY, PA, CT, VT, ME, NH; Midwest = OH, IL, IN, IO, MI; South = VA, NC, SC, FL, GA, AL.

operative ventures and mutual associations had distinct advantages over Bell, or even commercial independent carriers. Locating their systems in the center of farm areas, they eliminated the distance problem that afflicted Bell and other firms centered in urban places. By 1902 these noncommercial carriers were providing 60 percent of all rural telephones in the South. As even Bell managers came to realize, such local efforts were probably the only way of effectively serving many rural areas in the early twentieth century (see table 5.2).[47]

All together, commercial, mutual, and farmers' independent lines raised the number of rural telephones in the South to 10 percent of the total by 1902. Compared with the region's 87 percent rural population, the figure may not seem impressive (see table 5.3). But compared to Bell efforts before 1894, rural telephony had advanced considerably. Farmers, like all consumers of telephone service, gained from competition, which lowered prices and expanded output. More importantly for the farmer, it also allowed him a choice of service quality not available before.

Cases of successful competition such as these demonstrated that consumers could chose telephone service on the basis of price and service features in a competitive market, the beliefs of Bell managers to the contrary notwithstanding. As parent company managers slowly came to realize, independents made significant inroads throughout the United States wherever consumers felt that Bell service did not suit their needs. The nonurban Midwest, for example, experienced patterns of competition very similar to those

Table 5.3. Rural Telephones, 1902

	% Rural Tel.	% Rural Pop.	Semiurb.
New England	13	47	25
Mid Atlantic	2.2	38	23
Midwest	23	63	25
South	10	87	11

Source: Bureau of the Census, *Telephones and Telegraphs, 1902*, table 38. Bureau of the Census, *Census of Population, 1900* (Washington, D.C., 1904), vol. 1, tables 27, 37.

Note: Semiurb. = percentage of population living in semiurban places, defined in 1900 census; New England = ME, NH, VT, MA, CT, RI; Mid Atlantic = NY, NJ, PA; Midwest = OH, IL, IN, MI, IO; South = VA, NC, SC, GA, FL, AL.

of the South. As in the South, independents rapidly reduced Bell's dominance of the market and brought down prices.[48] Styling themselves "home" telephone companies, the new entrants appealed to local sentiments against the eastern firm. More than four hundred new companies competed with Bell in Ohio, Indiana, Iowa, Illinois, and Michigan, a third of the nationwide total of new firms.[49] They controlled many towns in these states, as well as the majority of rural sectors, though few major cities.[50] As in the South, new midwestern firms grew by offering cheaper, lower-quality service, carving out an important niche for themselves in the telephone market.[51]

Though there were basic similarities in all areas of competition, the situation in the South and Midwest did differ in important ways. These differences reveal the special problems of the southern telephone industry. Midwestern telephone competitors, for instance, had certain advantages over those in the South that permitted them a far greater range of opportunities. They received capital, equipment, and technical support from nearby electrical manufacturers eager to enter the lucrative communications equipment supply field.[52] This assistance allowed them to circumvent Bell patents on supplementary equipment such as switchboards, cables, transmitters, and batteries. From manufacturers such as Stromberg-Carlson and Kellogg Switchboard and Supply, midwestern independents also received support for competitive ventures in cities such as Chicago, though these efforts generally failed.

Midwestern independents benefitted as well from problems in Bell operating company structure in their region. Attracted by the high profits of urban communications and believing it needed to control key juncture points to expand its long-distance system, Bell poured capital into major midwestern cities, but paid little attention to smaller towns and rural areas. Chicago, Cincinnati, St. Louis, and Pittsburgh were each run by a different Bell operating company. Still another firm, the Central Union Company, was responsible for the hinterland.[53] In the South, Southern Bell served both urban and hinterland markets. It could, therefore, draw on profits from cities to sustain price wars with independents in smaller places. Central Union, however, was cut off from urban resources and fell prey to aggressive independent firms.

Because of these conditions, midwestern independents proved more successful in certain types of competition with Bell than did their southern counterparts. Connections to electrical manufacturers and early efforts in urban telephony served them well in the long run, enabling them to construct their own regional systems in places between large cities and rural areas where Bell's long-distance network was not yet extensive.[54] By connecting with each other, midwestern companies extended service from "the eastern slope of the Rocky Mountains to the Atlantic Coast."[55] These efforts surpassed anything undertaken by southerners, even the aggressive promoters of North Carolina. As a result, midwestern firms were better able to stand up to Bell's competitive use of long-distance lines, which had proved disastrous for non-Bell firms in Alabama and Georgia.

To combat Bell's head start in long-distance telephony and its advantages in system engineering, midwestern promoters also formed an association. With strong support from non-Bell manufacturing concerns, the Interstate Independent Association helped to overcome the limitations small size imposed on many competing firms. The organization started a clearing house to handle toll receipts and pooled resources to build main trunk lines to big cities. It also tried to bring political pressure to bear to overcome Bell's dominance of the Illinois Tunnel Company, the body that controlled access to Chicago's underground conduits.[56] Bell had been successful in keeping the city's conduits closed to competing firms, shutting them out of the key juncture point for long-distance service in the Midwest. Ultimately, the association failed to crack Bell's monopoly in the city. But the effort, along with the association's success in disseminating information on technology, finance, and politics in telephony, reveals the aggressive and organized nature of competition in the Midwest.[57]

In rural telephony as well, midwestern independents performed better than their southern counterparts. Both the South and Midwest had large rural populations, and in both places new firms attempted to reach these customers, but midwesterners did much better. Sixty-three percent of the Midwest's population lived in places of less than 4,000 people, yet 23 percent of the region's telephones were in such places. By contrast, 87 percent of the South was rural, but the rural South had only 10 percent of the telephones (see table 5.3). Overall, the ratio of rural population to rural telephones was 65 for the Midwest and 90 for the South.[58] While a high proportion of rural residents seems to have slowed telephone diffusion in the South, it had if anything the opposite effect in the Midwest. Seventy-nine percent of Iowa's population was rural, but 48.6 percent of its citizens had telephones by 1902.[59] Through such efforts, farmers in the Midwest had proportionally more telephones than urban dwellers by 1920, a startling reversal of the situation in the South.[60]

Midwesterners also built more sophisticated rural systems than did southerners. Rural southern customers tended to rely on the simplest forms of telecommunications. Thirty-two percent of all southern rural telephones were on farmers' lines. These affairs, constructed by farmers themselves using the least expensive equipment available, generally were not hooked into an exchange. Only 17 percent of the midwestern rural telephones were connected in this fashion.[61] Midwestern farmers started mutual companies and used lines from commercial firms. In this way they built an intermediate technology, providing fewer connections than Bell's more advanced network, but more than those possible with individual farmers' lines.

Entrepreneurship of this sort gave the Midwest the highest number of telephones per capita in the nation by 1902. Heavily rural Iowa was the preeminent state in telephone distribution. In

Table 5.4. Commercial Systems, 1902

	Tel./1,000 Pop.	Stations/SB	Toll/St./Yr.	Local/St./Day
New England	27	191	77	5
Mid Atlantic	29	283	97	6
Midwest	50	223	36	7
South	11	174	29	9

Source: Bureau of the Census, *Telephones and Telegraphs,* 1902, tables 29, 31. Bureau of the Census, *Telephones and Telegraphs and Municipal Electric Fire–Alarm and Police–Patrol Signaling Systems, 1912* (Washington, D.C., 1915), table 5.

Note: Stations/SB = stations per switchboard; Toll/St./Yr. = toll calls per telephone per year; Local/St./Day = local calls per telephone per day (325 day year); New England = MA, VT, ME, NH; Mid Atlantic = NY, PA; Midwest = OH, IL, IN, MI, IO; South = VA, NC, SC, GA, FL, AL.

the South, despite competition, telephones per capita were still the lowest in the nation (see table 5.4). Overall, midwestern promoters took the opportunities afforded by competition to fashion more effective alternatives to Bell's system than did southerners.

Some of these differences in performance reflected the Midwest's higher income and more prosperous agricultural sector. In 1900 southern farm income was less than half that of the Midwest.[62] Another important difference between the southern and midwestern telephone industry, however, was in the way the technology of communications was used in the two regions. As noted above, telephone distribution was low in the South; so too were messages per capita. Given the low number of instruments, this is not surprising. Few places in the South were connected by telephone, so those who had telephones could only make a limited number of calls. What is surprising, however, is the high number of messages per telephone in the region, higher than anywhere else (see table 5.5).

Elsewhere, messages per telephone were clearly influenced by telephone distribution. The relationship between telephones per capita, messages per capita, and messages per telephone was roughly constant and positive in the Midwest and New England (see table 5.5). Only the South breaks this pattern. There, despite

Table 5.5. Telephone Use, Nonrural, 1902

	Tel/1,000	Talk/Tel.	Talk/Capita
New England	27	1,293	51
Mid Atlantic	29	1,415	57
Midwest	50	2,196	110
South	11	2,698	29
US	30	2,138	64

Source: Bureau of the Census, *Telephones, 1912*, table 5.

Note: Tel/1,000 = telephones per 1,000 population; Talk/Tel. = talks per telephone per year; Talk/Capita = messages per population. See table 5.3 for regional definitions.

the narrow distribution of instruments, each one was put to intensive use. To achieve this high level of messages per telephone, either each southern telephone subscriber made many calls to a few other parties, or many nonsubscribers used existing telephones. With flat telephone rates, I suspect the latter was the case. The marginal cost of each call was effectively zero. At the same time, flat rates tended to restrict the distribution of telephones, as many southerners did not want to pay for their own telephone at prevailing prices.[63] This same price structure existed elsewhere, of course, but only in the South did customers respond by extensively sharing telephones, leading to the region's high number of messages per telephone.

Apparently, the sociology of communications permitted this unique southern response. The great increase in urbanization in the South between 1860 and 1900 had not, as some have argued, transformed southern society.[64] While New South towns were larger than the rural hamlets of earlier generations, more functionally diverse, more willing and able to attract industry and build services, they were still relatively isolated, homogeneous places. In contrast to the Midwest, moreover, southern towns and rural regions were not tightly integrated. Census statistics indicate that 25 percent of the Midwest's population lived in semiurban places, compared to only 11 percent for the South (see table 5.3). Semiurban is defined by the 1900 census as an incorporated place of less than 4,000 people. These intermediate towns probably stimulated interaction

Table 5.6. INDEPENDENT AND BELL TELEPHONE USE COMPARED, NONRURAL, 1912

Region	Bell Talk/Tel.	Ind. Talk/Tel.
New England	1,394	943
Mid Atlantic	1,244	1,469
Midwest	2,052	1,420
South	2,508	985
US	1,795	1,263

Source: Bureau of the Census, *Telephones, 1912*, table 25.

Note: Bell/Talk Tel. = talks per Bell telephone per year; Ind. Talk/Tel. = talks per independent telephone per year. See table 5.3 for regional definitions.

between city and hinterland in the Midwest, spreading use of the telephone. Most of the South's population, by contrast, lived in strictly rural places. New South cities, few in number and remote from the region's majority rural population, remained cut off from the large countryside economy.

Under these conditions, networks of communications in southern towns remained small and simple. The town drugstore, bank, or mercantile establishment served as the single center of information exchange. A few strategically placed instruments, used intensively, could serve the needs of an entire community, making the telephone in the South more a social than a private technology. The way in which independent telephones were used confirms this impression (see table 5.6).[65] Independent telephones were cheaper than Bell's, and probably more widely dispersed. At the same time, they were generally located in smaller, less densely populated places. More widely available and not located in strategic central places, independent telephones were not used in the same intensive fashion as Bell's.

This structure of demand helps to explains why, despite competition and falling prices, the southern telephone industry remained so far behind that of other places. Southern telephone needs could apparently be met by a simpler type of service, simpler even than that offered by independent firms in the Midwest. Southern town residents were satisfied with a few, centrally located telephones; rural residents opted for the cheapest, least sophisticated service. These demands for simple, inexpensive service, reinforced by southern poverty and industrial underdevelopment, formed the market to which southern entrepreneurs responded. Under these conditions, they did not have the same opportunities to build communications systems of the size and scope of their midwestern counterparts.

The southern response to opportunities in the telephone industry in this period also helps to clarify the nature of entrepreneurship in the region. In the initial years when the telephone industry depended on the vision and creativity of the entrepreneur, southern businessmen had been unable to meet the challenge. By 1894, when telephone technology, at least in its more rudimentary forms, was well understood, entry into the industry no longer required innovative abilities, only the capacity to imitate well-understood behavior and respond to immediate market signals. This imitative entrepreneurship southerners were capable of providing. They merely had to respond to the demands of the market, which was calling for less expensive, more widely available, lower quality, local service. Meeting such demands did not require large sums of capital or close connections with another, distant organization such as the parent Bell Company. Like textile mills, independent telephone exchanges became a part of town building in the New South, and thus fit the local orientation of most southern businessmen. By successfully tailoring telephone service to the specific needs of their area, southern independents, like those in other parts of the nation, fulfilled demands that Bell had missed or ignored.

Seen in this way the telephone industry of the South could be considered exceedingly efficient, rather than problematic. This perspective is too narrow, however, for while the resurgence of competition brought some short-term gains in efficiency, it also threatened other desirable long-term social ends. By the early twentieth century the South was developing a dual structure in telecommunications. Bell continued to build its system as best it could, but much of the South was served by small, remote, generally local firms offering a variety of types and quality of telephone service. While excellent for short-term, allocative efficiency, this structure threatened to leave large parts of the South outside of the evolving national communications network. At the same time, southern independent efforts were limited largely to the region's small urban sector; telephone promoters in the South rarely ventured beyond their city limits into rural areas, as their counterparts in the Midwest did. The potential long-term consequences of this situation were further isolation for the South and more missed opportunities in diffusing new technology.

No individual, group, or agency had either the authority or foresight to weigh these trade-offs in the nineteenth century South, or indeed in any other region. Even the national Bell Company was restricted in its options. To the company, the new carriers constituted a challenge to its plans for an integrated system. The corporation's primary interest was regaining market control, not improving the economy of the South. In addition, Bell's responses to competition were constrained by its inability to shift quality to meet consumer needs. As C. Jay French of American Bell recognized, it was difficult to tailor service quality to individual customers' preferences.[66] Doing so, moreover, would undermine plans for extending the long-distance network, which required high-quality equipment and engineering.[67] AT&T studies had already made it clear that building an integrated system required complicated feats of engineering and superior equipment, even at the local level. Profitable service depended on local feeder lines from regional companies, lines which had to be built of expensive metallic circuits.[68] Prices could not be cut too low if this system was to be profitable.

Competition had revealed a basic flaw in Bell Company policy. The corporation was seeking to integrate consumers with different preferences into a single, unified system. A large number of these customers were not prepared to join this system, at least not on the terms Bell was offering. Indeed, even if the national corporation had supplied the optimal level of quality nationwide, it still would have faced this problem in many parts of the country. In the South and elsewhere after 1894, many consumers still valued quality less than the average. They doubtless would have challenged Bell's planned system no matter how well conceived it was.

Adjusting to this new reality proved difficult for Bell managers. Plans for the Bell system had been laid early, and years of monopoly had produced in the corporation a deep commitment to a particular technological and managerial style. Changing that style provoked resistance, despite the presence of competition. Still, how Bell responded to this challenge could have significant ramifications, for the South and indeed the entire nation. If the firm abandoned its commitment to a single integrated system, then dualism in the telephone industry would remain and the South would continue in isolation. On the other hand, if the firm could find a way to mend the flaw in its approach and still pursue its goal of a national network, the South might gain long-term benefits. Throughout the first decade of the twentieth century, Bell personnel pondered their situation and groped toward a solution to the problems raised by competition.

Chapter 6

The Extension of the Network

A CAUTIOUS GAMBLER might have bet against the Bell Company after the first decade of competition. Far from declining, independents actually grew and increased their market share between 1900 and 1907. By that latter date they had almost one-half of the market nationally (see table 6.1). The South and Midwest, still leading in competing firms, moved farther out of Bell's orbit. Yet by the end of the decade, Bell was able to bring the industry back under its control and was beginning to extend its system into new regions. This change reshaped the southern telephone industry and gave the United States the modern "Bell System," which remained in place until 1984.

Important as Bell's victory was, it has never been fully explored. Some telephone industry scholars interpret Bell's triumph as striking proof of the superiority of its high-quality network. Noting the increase in toll facilities and long-distance messages, and the concomitant decline of independent market share, they assert that after 1907, the Bell system finally won public approval, ending the appeal of lower quality competitive service.[1] One version of this argument runs as follows: Public demand for long-distance service increased around the turn of the century; Bell made important technological improvements in long-distance transmission that independents could not duplicate. At the same time, the firm refused to permit independents to interconnect with its extensive system,

Table 6.1. PERCENTAGE OF INDEPENDENT AND BELL TELEPHONES BY STATE, 1907, 1912

Region	%Bell 1912	%Bell 1907	%Ind 1912	%Ind 1907
New England				
ME	76.3	69.8	23.7	30.2
NH	83.4	77.6	16.6	22.4
VT	59.3	54.9	40.7	45.1
MA	98.5	97.5	1.5	2.5
CT & RI	99.1	97.7	0.9	2.3
Mid Atlantic				
NY	83.4	73.6	16.6	26.4
NJ	86.8	83.6	13.2	16.4
PA	61.4	61.2	38.6	38.8
Midwest				
OH	39.8	37	60.2	63
IN	33.3	25	66.7	75
IL	58.7	48.9	41.3	51.1
MI	57.8	49.5	42.2	50.5
IO	30	15.9	69.3	84.1
South				
VA	62.3	56.6	37.7	43.3
NC	42.9	44.5	57.1	55.5
SC	63	62.4	37	37.6
GA	66.4	57.9	33.6	42.1
FL	41.8	36.3	58.2	63.7
AL	69.7	63	30.3	37
US	58.3	51.2	41.7	48.8

Source: Bureau of the Census, *Telephones, 1912*, table 24.

isolating them from the increasingly valuable toll network. Over time, the disadvantaged competitors lost the field, as the natural monopoly conditions of telephony become manifest.

Though not totally false, this interpretation misses many crucial aspects of the story. Competition had demonstrated that telephone service was not a natural monopoly. It had neither the declining long-run average costs nor the high fixed capital investment that restricted entry in other industries like electric power. A small-scale telephone independent with a relatively modest initial outlay could offer rudimentary but profitable service in even poor regions like the South.[2] Though Bell, with its more extensive system, could provide a level and quality of service that small-scale competitors could not match, that service was also more costly than independent alternatives. In places like the South, where demand for high-quality long-distance service was moderate, Bell's superior quality service was not a decisive factor in competition. Finding numerous customers who preferred their cheaper alternatives, independents flourished even while Bell built its integrated local and long-distance system. In the Midwest, competitors even began to make inroads in the long-distance market, suggesting that AT&T might not remain dominant even here.[3]

Independents in the South and Midwest had sold customers on their brand of service. The new firms were able to provide the inexpensive local service and limited regional toll connections many subscribers wanted. Satisfied with cheaper, if more limited non-Bell service, these customers were not prepared to switch back to Bell. Nor could the national firm easily convince them to do so. Bell's many improvements in long-distance service held only limited appeal to such people. The loading coil, for example, reduced costs and increased transmission quality on very long-distance lines, improving communications between New York and other major cities. These routes, however, had always been in Bell's command. The firm had always been strong, and independents weak, in those places which had a keen demand for long-distance service. The South, however, was not one of those places.

Technological improvements that aided Bell at its source of strength could not have been responsible for the firm's victories in the rural and small town markets of the South and Midwest. Independents in those places did not lose the field because they could not duplicate Bell's long-distance service. Had they been competing directly with AT&T over its major trunk routes, they would have folded quickly. Their lack of capital and technical skill, and their late start would have proved insurmountable disadvantages. But in a market divided by quality preferences, independents did not need to compete in this fashion. Independent facilities did not require large amounts of capital or engineering skill, as did long-haul trunk lines. So long as independents did not need to construct an integrated national system, they held a price advantage over Bell. They continued to compete successfully by offering a different type of product at a lower price. Under these conditions, neither Bell's supply-side improvements nor its refusal to allow independents into its network could have been responsible for its competitive triumph.[4]

In the South and similar regions, a new entrepreneurial strategy, not technology, brought about Bell's resurgence. As Southern Bell managers realized, competition had revealed the severe limitations of the Bell system in marginal economies such as the South. Southern Bell responded to this situation in an innovative fashion, devising a means of retaking the southern market while fulfilling the goals for system integration set by parent company executives. The company came up with a new marketing strategy which made the Bell system appealing to consumers of independent service. At the same time, it made key compromises on quality and price that still left the goal of system integration intact.

These important innovations proceeded slowly, however, despite the continued growth of competition in the early twentieth century. The new policies, which took account of regional variations in supply and demand conditions, ran counter to the conventional wisdom of parent company personnel. Managers at American Bell maintained that local needs had to give way to Bell system standards, and tried, uncompromisingly, to bend consumer demands to their system. Regional variations in economic conditions, they held, should not stop system growth. E. J. Hall, ironically one of the chief architects of the new strategy, had articulated these conventional assumptions in the 1880s. He had pleaded with Bell licensees to take account of the needs of the long-distance system and adjust local operations accordingly.[5]

After 1894, other Bell personnel tried to reformulate Hall's ideas to fit the new competitive environment. As company attorney George Leverett succinctly put the matter to AT&T president F. P. Fish: "I take it that it is extremely important that we should control the whole toll-line system of intercommunication. . . . Such lines may be regarded as the nerves of our whole system. We need not fear the opposition in a single place, provided we control the means of communication with other places."[6] Thomas Doolittle, an AT&T engineer and early builder of long-distance lines, bolstered Leverett's pronouncement with a thorough analysis of the Bell system's effectiveness under competition. Doolittle maintained that long-distance development, coordinated from the top down, could serve as the company's first and foremost weapon against opponents. Most of Bell's competitive problems, he maintained, stemmed from operating companies' underinvestment in toll facilities. If the parent company would pour capital into this end of the business, Bell could quickly surpass its rivals and extend its system.[7]

Doolittle's recommendations followed from observations he had made in the 1890s. Within any one town or exchange area, he noted, telephone use grew in more or less constant proportion to the number of places connected, as each customer made a constant number of calls.[8] Beyond a certain point, increased local connections added little to the utility of the telephone to each subscriber. Interexchange connections, however, greatly augmented the telephone's value, especially to important subscriber classes such as businessmen. Under these conditions, demand for higher quality service, the kind Bell was prepared to provide, would follow automatically from increased long-distance facilities.

From this key observation, Doolittle averred that system interdependence was an inescapable fact of the telephone business. As he put it, "the toll line business is governed by laws as immutable as the laws of nature."[9] Invariably, the toll system would expand and envelop local systems. If so, then AT&T had little to fear over the long run from its competitors. Once local and long-distance service were joined more fully, telephone demand would increase geometrically. The firm that had the most extensive system—clearly Bell—would gain the most customers.[10]

As even Doolittle realized, his recommendations raised new problems that would have to be solved before Bell would have its victory. Operating companies, for example, projected only expected demand for local exchange service. This approach underestimated the needs of an integrated system and led companies to set investment levels too low. Low investment at the local level led to inadequate capacity for the whole system.[11] Users of long-distance lines paid premium prices for their circuits, and they expected premium service. If they could not get it, they would not opt for long-distance service, undercutting system integration.

To combat this danger, Doolittle tried to instruct the regional firms in the correct methods of operation. To overcome the planning problem, he advised operating companies to employ earnings per capita as their index of required investment, not the more common earnings per subscriber. The latter was inaccurate because it did not take account of the great increase in subscribership possible with an integrated system. Doolittle also listed four principles for operating companies to follow.[12] They should aggressively seek new customers for local service; extend toll service to new places as quickly as possible; use metallic circuits to raise transmission quality; and employ improved methods for handling toll calls. If applied together, these principles would increase local company revenue and pave the way for system growth. Just to be sure, however, Doolittle also recommended that AT&T coordinate operations to prevent insufficient capacity, poor equipment, and high costs at the local level from subverting long-distance service.

Important as Doolittle's work was, it was not designed to deal with the problems present in regions like the South. All of his suggestions rested on his belief in the inevitability of system interdependence. Without that faith, he could not guarantee that further investments in long-distance construction would pay off in increased market share for Bell companies. The engineer justified his faith by studies of Bell operations in New York, New Jersey, and New England.[13] But these studies had only limited application in areas of heavy competition such as the South and West. Doolittle's own observation of Missouri and Kansas Bell suggested that his recommendations might not work under certain conditions.[14] Indeed, though he did not realize it, they were uncertain at best in regions which did not conform to northern conditions. Yet these other regions were the proving grounds of his policies; if Doolittle's suggestions could not help to defeat competition in its strongholds, they had little to offer.

When empirical studies failed him, Doolittle made yet another assumption. Even in places of heavy competition, he maintained, the public would eventually become accustomed to long-distance service and switch to Bell as new facilities were built. On this basis, Doolittle advised operating companies to reject strict profitability as their index for construction requirements. Some facilities would have to be built ahead of demand, on the hope of future profits.[15] But future profits were cold comfort to operating company presidents who had to worry about their firms' immediate financial health. Their capital resources were limited and competition had cut deeply into their market share, forcing them to pay strict attention to immediate profits.[16] Recommendations for massive expenditures in the face of falling prices and profits seemed sheer madness to beleaguered regional managers.

These men realized, as Doolittle did not, that competitors fared quite well by offering cheaper, more limited service. An AT&T trunk line from Atlanta to Washington might lure some customers away from the competition, but there was no guarantee that enough would switch to pay for this undertaking. A well-rounded toll network might help against independent carriers in the long run, but the question remained—how to construct such a network in the face of ruinous competition. Faced with this harsh reality, operating companies such as Southern Bell instead formulated their own responses to competition, responses quite different from Thomas Doolittle's.

Southern Bell and other companies began by fighting competitors on their own terms, lowering service quality and reducing prices.[17] Immediately following competition, Southern Bell greatly reduced its rates, introduced new classes of inexpensive service, and provided more flexible terms of payment to match the independents.[18] In Atlanta, which had a first-class exchange, annual prices had been as high as $100 for business and $70 for residential service.[19] They tumbled 25 percent in 1894.[20] In other cities reductions were even steeper, as the company approved across-the-board cuts to get prices down to "fighting levels."

Doolittle and others in the parent company objected to cut rates, arguing that the Bell system demanded higher charges commensurate with the quality of its service. Partly in response to these pleas, Southern Bell rescinded the across-the-board reductions in 1895.[21] Nonetheless in places of fierce competition discounts were hard to avoid, and in some cases the company offered free service to eradicate its competitors.[22] In other cases, recognizing that standard Bell exchange equipment was simply too expensive to fare well against competitors' cheaper models, SBT&T managers asked for permission to build small magneto exchanges for thirty or forty subscribers.[23] For those who could not afford standard exchange service, the southern company also offered cheap, multiparty lines to match those of its competitors.[24]

As Doolittle realized, his policies could not be enacted if regional companies responded to competition in this fashion. Reducing the quality of local equipment would make interconnection with long-distance circuits more difficult, while slashing prices drained the revenues needed to undertake expensive new construction.[25] Doolittle proposed two innovations to deal with the realities of competition and make it easier for operating companies to follow his recommendations. First, he proposed a new marketing strategy to convince customers that the Bell system was a better choice than independent alternatives. Second, he drew up a new Bell organizational structure, to allow operating companies to push the Bell product into the competitive margin without going bankrupt.

To "sell" the Bell system, Doolittle recommended that operating companies employ several new types of small town service in competitive areas. On his suggestion, American Bell made class F party lines available to connect rural customers and other outlying residents in sparsely settled regions to nearby Bell exchanges.[26] As an intermediate step between full-fledged class A exchanges and such simple systems, Doolittle also developed a new type of service for what he termed "Petersham" towns.[27] In such places, too small for exchange service, Bell installed public toll stations. In this way customers could learn the value of long-distance service. This lesson would prepare them, Doolittle hoped, for a Bell exchange: demand for long-distance service would thus stimulate demand for local service.

Doolittle also proposed that AT&T assist local companies by extending its long-haul trunk lines through major metropolises in troubled regions—Cleveland, Chicago, St. Louis in the West; Richmond, Atlanta, Birmingham in the South.[28] Believing such progress would stimulate consumer demand for long-distance service, Doolittle also hoped it would convince operating companies to undertake toll line construction themselves.[29] Forced by increasing customer demand to upgrade and extend their feeder toll lines to connect with AT&T trunks, Bell operating companies would have to make the expenditures the system required. AT&T and operating company toll lines could then meet in an expanding Bell network. In the end, such efforts would surround independent firms and "leave the burden of maintaining unprofitable plants in the hands of local [non Bell] people."[30]

Though steps in the right direction, these recommendations still fell far short of what was needed. For all their originality, they too rested on a faith in the inevitable triumph of the Bell system. Doolittle's proposed cheaper facilities, for example, still cost more than independent alternatives.[31] The Petersham town, an interesting concept, was actually a very simple marketing device. Based on a learning-by-doing model, it implied that customers would autonomously come to prefer Bell service over independent alternatives. But if, as many operating company personnel were beginning to believe, many customers had turned to independents because they offered a preferable type of service, then Doolittle's innovations could not succeed. Under certain favorable circumstances, customers might switch to Bell's superior quality system, but the key was to sell that system in places where conditions were not yet favorable.

When confronted with these limitations, Doolittle counseled faith and continued to implore Bell companies to increase toll line expenditures. At one point, for example, he held that if Bell firms merely established a monopoly in town after town, they could use the monopoly profits to make necessary expenditures. This proposition, of course, begged the question of how the firms could reestablish a monopoly when customers were apparently satisfied with service from competitors.

Faced with operating company opposition, Doolittle came to believe that his policies were sound, but Bell's organizational structure was faulty. He maintained that the root of operating company resistance to his plans was the "irrational" and "idiosyncratic" organization of the Bell companies themselves. All problems flowed from the original division of the Bell system into parent and operating units, and the organization of operating units along territorial lines. If AT&T were to reorganize all operations along functional lines, the engineer maintained, it could eliminate most of its problems.

Studies of Bell's organization seem to confirm this belief. Examining company organization in the Midwest, Doolittle concluded that Bell's territorial structure inhibited necessary toll line construction. Though he believed that the managers of the midwestern Central Union Company were "unusually capable," he found them inexplicably unwilling to make the necessary 2.3 to 3 million dollar investment his strategy called for.[32] The problem, Doolittle reasoned, was that Bell had left the major midwestern commercial centers in the hands of several different firms.[33] In this situation, the managers of Central Union "[were] very apt to fix an imaginary center in [their] territory, overlooking the requirements of adjoining territory."[34] As a result, they tended to discount the value of toll lines whose main trunk joined two cities outside their territory, but which nonetheless could be of benefit to all. By ignoring such "extraterritorial" connections, operating companies underinvested in local and long-distance plant, resulting in the dreaded problem of inadequate capacity, which strangled system growth.[35]

These difficulties AT&T management termed the "border problem."[36] It occurred wherever an operating company had to build lines that terminated in an adjacent company's territory; or where a firm had to build an intermediate link on a line which began and ended in a territory other than its own. Theoretically, cooperation and profit sharing between firms could alleviate this problem. AT&T provided extraterritorial contracts for this purpose, but Doolittle still maintained that AT&T's decentralized structure prevented a successful resolution.[37] To the engineer, the "imaginary center" that regional firms' managers fixed in their own territory was an insurmountable obstacle. Doolittle believed that so long as Bell continued to rely on a territorial division of responsibility, the problem would persist. So long as it persisted, local companies would continue to underestimate the need for construction, depriving AT&T trunks of adequate feeders and starving the system.

The only answer to this dangerous situation, Doolittle held, was greater top down coordination by AT&T.[38] His organizational scheme called for AT&T to "rationalize" the industry by eliminating territorial units and drawing together parent and operating companies in one grand organization. AT&T could then provide needed leadership in the field, carrying out the toll construction that operating companies themselves refused to undertake. The long-distance company could also apply the principles of efficient design, planning and routing at all levels of operation. This structure would allow Bell's toll network to expand and engulf the competitors.

The solution to Bell's problems seemed simple to Doolittle. Regional variations in economic conditions could not stop system growth if Bell merely did what it did best—continued to build its high-quality long-distance network and integrate it with local service. Doolittle's belief, however, rested on his assumption that Bell had a type of service the public wanted and that its rivals could not provide. As we have seen, long-distance service did give Bell a competitive advantage in certain relatively developed parts of the South and Midwest. Yet in many other sectors, independents remained firmly entrenched. Bell had denied competitors access to its long lines, locked them out of key points such as New York, Chicago, and St. Louis, and used profits from monopolized intercity markets to subsidize price wars in competitive local ones. Still the independents grew. Where the challengers could not breech Bell's competitive walls, they turned elsewhere in the still largely open telephone market. The result was continued competition and further erosion of Bell's market position.

For Doolittle's policy to succeed, customers in America's vast rural and hinterland areas would have to become convinced that the Bell system was more attractive than cheaper independent alternatives. But Doolittle did not offer an effective marketing method to allow Bell to overcome customer resistance. His organizational schemes suffered from a similar myopia. In 1907 AT&T did reorganize operations along functional lines, though it retained some of its decentralized structure and territorial divisions. While Doolittle had foreseen that Bell's competitive triumph depended on reorganization, he had overestimated the degree of change necessary. The trick he had missed was combining overall coordination and control with local flexibility.

In the short run, Doolittle's efforts tended to reinforce Bell managers' erroneous belief that the triumph of their system was determined by technology. Over the longer term, however, Doolittle's failures proved to be an important part of the innovative process, opening up lines of thought that later innovators would pursue with success. Other members of the Bell system, working in the South and Midwest, grasped what needed to be done. Rejecting Doolittle's faith in the Bell system's inevitable triumph, they abandoned his dream of central planning. Planning appealed to the mind of the engineer, but it could not deal with the very real economic differences between and within regions. Recognizing that these differences were the product of consumers' preferences, not irrational organization, they pioneered the new policies that made the long-awaited Bell system a reality.

In 1896, AT&T took an important step in this direction when it sent E. J. Hall, the firm's open-minded vice president, to head Southern Bell. Hall was a man of many talents. In the 1880s he had outlined one of the first plans for the long-distance network. Later, in Buffalo, he had served as president of another operating company and molded an exceptionally fine firm that stood as a model for other Bell companies.[39] To this blend of experiences he added a deep understanding of the relationship between technology and organization. He also proved a quick student of finance and an innovator in this field as well during his tenure at SBT&T. All of these skills served him well in the South, where he faced a situation quite different from his previous experiences in the North.

In his first report, the new chief executive acknowledged the difficult fight ahead.[40] Despite efforts to avoid reducing rates, competition had driven Southern Bell prices down to what were becoming unprofitable levels. In Concord, North Carolina, annual subscriptions were cut to $10; in Charleston, South Carolina, competition had forced the company to offer a 50 percent discount to some subscribers.[41] Even in the numerous towns where Southern Bell faced no direct challenge, the threat of entry had forced it to provide discounts and reduced rate service. Hall moved quickly to staunch the loss of income by limiting price competition as much as possible. Though he realized that prices would never climb back to precompetitive levels, he also knew that Bell's higher quality system required higher rates. Cut rates, he wrote to AT&T president Hudson, would not provide enough funds to pay the "system costs" that were a part of Bell service.[42]

Borrowing a page from Doolittle's book, Hall turned to improved service quality and system integration as his chief competitive tactic. Between 1897 and 1908 he upgraded the southern toll network to make it more compatible with AT&T equipment, investing considerable funds in toll line construction and in switching exchange circuits from grounded to metallic.[43] Though acknowledging that "it is certain that the population of the South is much inferior in character, capital, and energy" to the rest of the nation, he believed that places of sufficient telephone demand existed in the region to allow the Bell system to expand profitably.[44] According to his own estimates, exchange revenue per station was higher in the South than in the Central Union, or New York and Pennsylvania companies' territories—both places of heavy competition. The difference in income between these two regions and the South lay in toll revenue per station, only $.28 in the South, compared to $.77 in Central Union and $1.03 in New York and Pennsylvania. Even the Southwestern territory adjacent to Southern Bell had a higher toll revenue rate. These statistics, Hall reasoned, indicated that Southern Bell had underinvested in toll lines, a situation he proposed to reverse by raising average southern expenditure per station from its low of $116.00 to a level comparable with other territories, approximately $150.00.[45]

Unlike Thomas Doolittle, however, Hall quickly realized that these bold plans required important changes in management and finance. The new investments were difficult for SBT&T to fund given its present state.[46] As of 1897, the company had issued all of its authorized capital stock, $1,000,000. Increased floating debt, the president maintained, would be neither wise nor sufficient. Yet few alternatives existed. Southern Bell was woefully undercapitalized. In 1889 it had relinquished 35 percent of its stock to ABT in payment for a permanent franchise. This move had raised the firm's liabilities to $725,000, but added no new funds to the business. SBT&T financed subsequent expansion out of retained earnings; in 1889 it gave stockholders 2,000 more shares of stock in lieu of a hefty dividend.[47] Rejecting this conservative financial policy, Hall proposed instead to increase Southern Bell's authorized capital stock.

This recommendation met with adamant resistance from cautious stockholders, especially Western Union. Western Union had provided conservative leadership to Southern Bell during the monopoly period. As we have seen, its policies made Southern Bell unwilling to undertake long-range construction and shouldered it with a preference for inexpensive equipment. Under competition, Southern Bell's caution increased still more, as Western Union and the other stockholders had neither the resources nor the vision to embark on expensive projects in the face of declining profits. Even after American Bell took majority ownership in Southern Bell, Western Union retained a substantial minority interest. It feared that further issues of equity capital would erode its shaky position in the firm.[48] AT&T, with its superior resources, would be able to take the lion's share of new securities, all but eliminating Western Union's voice in management.

To counter such opposition, Hall resorted to a subterfuge. Against his true wishes, he borrowed substantial amounts of cash from American Bell at two-week intervals, until Southern Bell's accounts payable increased by $527,000 in 1897.[49] Knowing that this method of finance could not continue, he again pleaded for an increase in the company's capital stock.[50] Still the stockholders refused to budge. Accounts payable increased by another $469,000 the next year, prompting Hall to remind the owners that their firm's floating debt now exceeded the amount authorized in its charter by $214,111.68.[51] Willy-nilly Hall proceeded with his plans, as he and the other officers solicited loans from the parent company and anyone else willing to extend credit.[52] Even in 1899, however, when Southern Bell's short-term debt shot up by another $740,000, the stockholders did not take action.[53] By 1904 SBT&T's balance sheet showed debt of $11,469,618, and a capital stock of still only $1,000,000.[54] Finally the stockholders increased their firm's capital to $30,000,000. Between 1905 and 1906, Southern Bell issued $15,000,000 in new stock to AT&T to pay part of the outstanding debt, reducing it to $1,000,000 by 1908.[55]

By aggressively pursuing expensive construction and redevelopment plans, Hall had forced Southern Bell to increase its long-term liabilities. AT&T played a major part in this effort by assuming SBT&T's short-term notes and converting them into stock. Beginning in 1901, the long-distance company entered into a financial partnership with Southern Bell, providing in that year $250,000 for toll line construction and switchboard improvements, and later another injection of $750,000 to continue the work.[56] Hall also used AT&T resources to buy out minority stockholders, fearing that they might object to his future plans.[57] This work was completed in 1908 when Southern Bell gave over complete ownership to AT&T to pay off the last of its notes.[58] By refinancing the firm—increasing its capital resources thirty-fold in a decade—and tightening its links to AT&T, Hall overcame Southern Bell's long tradition of fiscal conservatism. The result was a new financial stability, which allowed Hall to resume the dividends he had stopped during the financial crises of the preceding years. Thereafter Southern Bell returned a steady 6 percent to its investors.[59]

The new president's financial policy was just the sort of innovation needed to make the strategy of system expansion effective in the South. Firmly convinced that he was on the right track, Hall wrote an optimistic report to AT&T president Fish in 1901. "The facts . . . show very clearly," he stated, "the importance of extending our toll facilities as a means of controlling the situation, in addition to the direct profits to be derived from their operation."[60] This analysis seemed to confirm much of what Doolittle and others had been saying—operating companies needed to make greater expenditures on toll line facilities and to improve local service to overcome the independent challenge. But Hall, almost alone at this time, also understood that this policy by itself would not guarantee a complete victory. It would have to be just one part of a more comprehensive strategy.

As Hall quickly perceived, while toll lines proved effective against independents in the more developed parts of the South, other areas, such as the numerous rural counties, stuck with the competitors.[61] Frustrated by these experiences, the Southern Bell president reformulated his competitive strategy, reworking Doolittle's policies in a way which opened up new options for dealing with competition, particularly the harsh reality of continued customer preferences for independent service. Combining his older working knowledge of telephone operations with his recent experiences in the South, Hall produced the most important innovations in the early twentieth century telephone business.

Hall's first breakthrough was a new entrepreneurial strategy that dealt with demand side limits on Bell system expansion. This strategy grew in part out of his work on finance. While Hall had been able to put Southern Bell back on a solid financial footing, he soon realized that continued financial health required a new approach to long line expenditure. Across-the-board toll line construction only increased costs without providing sufficient additional revenue to justify the expense. Hall maintained instead that the best plan would be to build up southern toll and trunk lines in areas such as Georgia, Alabama, and Virginia first, where customers had already shown an interest in long-distance communications and where profits were relatively high.[62] These profits in turn could be used to gradually acquire more territory for the southern toll network.[63]

Even while expanding the Bell system in the South, therefore, Hall began to modify the bold plans set forth by men such as Doolittle. Working aggressively in key states, the Southern Bell president concentrated capital in important cities and commercial centers. There he upgraded southern equipment, reconstructed old lines with metallic circuits, added new switchboards, and installed underground cables.[64] But he wisely did not attempt to fill in all the gaps of the southern network. Instead, focusing his resources where they would do the most good, he expanded this network gradually, grabbing territory at the competitive margin wherever he could while avoiding costly confrontations with independent firms. By looking more carefully at the return on individual projects, he kept Southern Bell financially strong, quelling fears that expansion would hurt the operating company's profits.

Making system development contingent on immediate profits and regional commercial conditions was a more restricted approach than Doolittle's, but it was also more realistic. Recognizing that customers did choose service on the basis of quality, Hall took account of an important demand-side constraint. In doing so, he checked his firm's loss of market share, as Bell and its rivals divided the market, each serving the territory they served best. But this policy did not completely satisfy those members of the parent organization—Hall included—who continued to entertain visions of a national, integrated network under Bell control. The competitive equilibrium Hall had established in the South threatened to leave large portions of the telephone market out of Bell's reach. Accordingly, Hall sought some means of extending the Bell system to those places still committed to independent service.

The most likely and fruitful territory for further growth was the large sector between city and farm. In this territory of midsized commercial towns and progressive upcountry cities, the market for telephone service was volatile and quixotic. Customers wavered between Bell and independent options, as the rivals competed vigorously for market share. In this competitive margin, the choice was not cut and dried. Aggressive local entrepreneurs fended off Bell incursions by offering limited toll connections; Bell captured territory by gradually building its system out from key points. The deciding vote as to which option such towns would choose often rested with the local business community. They not only made up the preponderance of telephone customers, but they held both formal and informal power in city government; their choice of service often decided the issue for the entire town. Recognizing that local businessmen wanted and valued outside connections, SBT&T made a special effort to draw them into the Bell system. The company strove to beat the independent firms by providing these customers with more outside connections and offering them special deals such as a credit for long line service equal to the yearly exchange rate.[65]

As an adjunct to this strategy, Hall also began to make

acquisitions of rival firms at key points. Expansion and acquisition had to work in tandem, for purchases could only be profitably made in places that had some demand for Bell-type service. Carefully weaving his way between the need for financial stability and the desire for Bell system expansion, Hall deftly acquired a stake in a number of important independent telephone companies.

Acquisitions often required protracted negotiations with towns to bring them under control at a minimum cost. In Richmond, for example, Hall worked tenaciously to acquire a nettlesome rival. The business elite of the city, supporting a competing firm, had in the 1890s engaged in a bitter legal struggle with Southern Bell for an exclusive right-of-way franchise. The local interests triumphed in 1897, temporarily ousting Bell from the city. But apparently the legal machinations had only been a ploy to drive up the value of the independent property. In 1898 the owners offered to sell out for $310,000.[66] The competing company was almost wholly supported by Warner Moore, a Virginia businessman, who had supplied $93,000 in bonded capital and owned a significant block of stock in the firm. Moore was not a telephone manager, however, and fearing damage to his property from renewed competition, wanted to get his money out quickly.[67] While Hall was not willing to pay this "wildly absurd" figure, he did not close the door on negotiations altogether. Using time to his advantage, he waited until 1901 and arranged more favorable terms.[68] The local investors, content with a modest profit for their property, happily dropped all pretense of "antimonopoly" rhetoric and embraced the "foreign" Bell Company.

In Richmond, Southern Bell operated from a position of strength, as local entrepreneurs could not meet the city's heavy demands for high-quality service and extended toll facilities. In smaller places, acquisitions required a finer hand. Through careful compromises with local elites, Southern Bell gained control of Washington, Georgia, a small but vital town which fed into the important Atlanta-Augusta long line. The opposition firm in the city had gone bankrupt, but Southern Bell was constrained from acquiring its equipment by the small size of the town's market.[69] Southern Bell managers W. T. Gentry and James Easterlin, however, believed the town could support a class E exchange of thirty subscribers. While this venture would not be very profitable, it might pay off if a branch line from Washington to the Atlanta-Augusta trunk could be arranged. As Easterlin put it, "if we do not build the toll line, it will be hardly possible to build the exchange."[70] To save valuable funds, the Southern Bell managers contacted local businessmen, who agreed to finance the line and patronize the small Bell exchange exclusively, eliminating the threat of renewed competition.

W. T. Gentry, the long-time SBT&T executive, worked closely with friends in the town to make this arrangement possible. Having been born in the South and having served in the Georgia territory for many years, he was well respected by the state's businessmen. Through friends, Gentry arranged for the purchase of the old opposition exchange at auction for only $125.00. He then quickly installed a class E Bell exchange at the Hammack and Lucus drugstore. With assistance from local promoters, he built a toll circuit to Barnet, looping into the Atlanta long line. The compromise satisfied all parties: Bell gained control of a strategic point and Washington received the local and regional connections it wanted. Cooperation with local interests had provided a victory that Bell could not have achieved on its own.

In cases where such cooperative ventures could not be arranged, Hall purchased certain competing companies outright, if they served strategic links for the southern network. Fearing that competing toll companies would build their own independent systems, Hall bought a number of Georgia independents and joined them to Bell toll lines running west from Atlanta and north through Birmingham. He acquired firms in the towns of Valdosta, Gainesville, Waycross, Thomasville, Brunswick, Sandersville, Loganville, Lawrenceville, Tennille, and Jackson. These towns fed directly into their state's well-developed toll network.[71] Though not very profitable in and of themselves, their strategic position justified the purchase.

In rare cases, Hall also purchased competing firms in towns where Bell had no competitive advantage or evolving system. In Charlotte, North Carolina, for example, a number of local businessmen had built an effective independent exchange and had begun to provide limited regional toll connections as well. Charlotte was a growing New South city of 20,000 people and a commercial center for southern textile mills. It would, Hall knew, become a key long-distance point in the near future. Yet Bell had no system in the area comparable to that existing in Georgia and Alabama. Hall rightly hesitated in acquiring the opposition exchange in North Carolina, for such purchases could put a strain on his firm's capital resources.[72] Increased capitalization, he had already established, depended on the return to individual projects. Too aggressive an acquisition policy would undermine profits.

Unable to force a compromise with the independent entrepreneurs and convinced that Charlotte was too valuable an opportunity to pass up, Hall reluctantly advocated direct engagement. As he wrote to AT&T president Hudson, "While I do not believe in the policy of buying out the opposition companies generally, I think there are exceptional cases where it should be done and this is one of them."[73] The Southern Bell president relied directly on AT&T support, and, through surreptitious transactions that never appeared on Southern Bell's books, borrowed money from the parent firm to purchase the Charlotte property.[74]

In this instance, the future value of the city to the Bell toll network justified the acquisition. But in general, Hall carefully weighed each purchase. The decision to make a costly acquisition hinged on its potential value to the Southern network, and on the potential cost of increased competition if the purchase was not made. Under these criteria, acquisition, while a useful policy, clearly had its limits. Its effectiveness in places such as North Carolina, where Bell had yet to establish an extensive network, or smaller towns and rural sectors that stuck loyally to independent options, was severely curtailed. Indeed, even cities that, it would seem, should have been part of the Bell empire could remain independent if local promoters were aggressive enough and customers loyal enough to home-grown companies.

In Lynchburg, for example, Southern Bell was unable to buy out its rival despite repeated efforts. The town was a valuable point in the firm's growing Virginia toll network, providing feeder traffic to its trunk lines. Yet Virginia, with its relatively high income and level of development, had fostered a healthy and aggressive independent movement. As a result, Lynchburg continued to support both a Bell and independent exchange.[75] Financed by wealthy Virginia capitalist Charles Scott, the independent exchange had done well in attracting disgruntled Bell customers. By 1896 it had 500 subscribers and a reported income of $4,000 on a $17,000 investment. Though the firm's exchange rates actually exceeded Bell's, its subscribership grew to 730 by 1901. Under this rare head-to-head competition, Bell rates had fallen precipitously, leading to a rapid depreciation of its property and an operating deficit of $2,400 for 1896.[76] Despite strenuous negotiations, Hall declared in 1901 that "conditions there are not yet such as to make it likely that we could reach an agreement."[77]

In cases such as these, where acquisition was impossible or too costly, Hall relied on special contractual compromises to gain control of the market. This was the second part of his new entrepreneurial strategy. The agreements permitted him to expand the Bell system, but also allowed competitors a permanent place in the industry. Sublicensing was the most common form of contractual compromise. During the early years of telephony, Bell had authorized agents to license their own subordinate subagents. After 1880, with the formation of operating companies and the establishment of its monopoly, Bell abandoned this policy. Between 1881 and 1894, the firm strove for greater central control. But the reopening of competition forced some members of the Bell organization to reconsider sublicensing policy. Men such as E. J. Hall began to see that these contracts, which delegated partial responsibility for Bell telephone service to other parties, were useful in marginal areas that Bell itself could not profitably serve.

Sublicense contracts varied in specifics, though they all involved a degree of cooperation between Bell and non-Bell firms. Often Southern Bell took partial ownership of the connecting firms, extending funds so that they could upgrade equipment or purchasing their stock to provide financial stability.[78] Other times the erstwhile competitor agreed to divide responsibility with Bell for financing and operating a particular line.[79] In most cases, Bell ownership stopped short of total. The contracts were not designed to make independents Bell subsidiaries, but, in effect, licensees. As licensees, they were entitled to connect their local systems into Bell's expanding toll network.

Table 6.2. Southern Bell Sublicensed Stations

Year	Sublicensed Stations	Total Stations	% Sublicensed
1900	1,707	31,631	5.4
1901	3,094	34,615	8.9
1902	4,626	42,612	11
1903	n.a.	—	—
1904	n.a.	—	—
1905	19,306	88,159	22
1906	27,106	108,491	25
1907	35,375	120,023	29
1908	82,225	212,759	39
1909	108,863	262,604	41
1910	128,575	302,368	42
1911	146,644	339,542	43

Source: SBT&T *Annual Reports.*

Though Bell had the larger system, it found that even more than its smaller competitors, it benefitted from this sort of interconnection.[80] Dividing responsibility with competitors in this fashion enabled the company to achieve its goals at an affordable cost. By selling its property and equipment to independents in towns where they had proven themselves the better competitor and then executing sublicense contracts, Bell saved needed capital yet still retained a degree of control over its rivals.[81] Such compromises ended the unrealistic dream of complete competitive victory pursued by men such as Thomas Doolittle. In remote parts of the South and similar regions, however, they also proved invaluable to Bell's continued growth.

Under Hall's direction, the percentage of sublicensees in the South increased dramatically, from 5.4 percent of Southern Bell's stations to 43 percent between 1900 and 1911 (see table 6.2). Throughout the South, Hall used the contracts to nullify competitive threats where other methods failed. In Asheville, North Carolina, for example, Southern Bell compromised with the aggressive Asheville Telephone Company. Bell supplied capital to build a toll line from Charlotte to Gastonia, but it left local service in the hands of the independent company.[82] By providing the necessary capital, SBT&T was assured that the Asheville firm would build its lines in accord with Bell system requirements; by retaining the home company's cooperation it muted potential conflicts with local business interests and indirectly assumed management of an important exchange.[83]

In a similar manner, Southern Bell made arrangements with merchants in Waynesboro and Mobile, who in 1897 were beginning to demand better toll connections. SBT&T was still in the midst of its financial crises and had "no money to build any of the lines wanted."[84] Fearing that if something was not done, local companies would begin to link independent exchanges on their own and circumvent the Bell system, Hall was willing to modify company policy and allowed an independent, the Southern States Telephone and Telegraph Company, to build lines under Bell supervision. Hall allowed Southern States Telephone to maintain its own exchanges and toll facilities and connect with Bell toll lines at certain key points, so long as the firm modified its equipment to be compatible with Bell's and rented only Western Electric instruments.[85] By using local resources, Bell met the demands of the town subscribers quickly and cheaply and prevented the independent from operating on its own.

Hall became a fervent supporter of compromise with independent companies where the only alternative was futile competition. This policy fit his competitive strategy perfectly. From his study of toll lines, he had realized that Bell did not, and indeed could not, operate every point in the South; but with its control of technology and capital, it had no need to do so. Sublicensing allowed Bell to operate indirectly in otherwise inaccessible places, while it prepared marginal markets for the switch from low- to high-quality service by introducing them to the benefits of the Bell system.[86] It was also politic in places like the South to preserve a role for local capital, for as Hall realized, people were much more willing to accept service from a "home" corporation than a "foreign" one, even if the home company was controlled indirectly by Bell.[87]

Still, Hall had to convince his superiors of sublicensing's value. In 1897, American Bell did not even have a standard contract to make the connections with the Southern States company that Hall desired.[88] Hall, in his pivotal role as SBT&T president and AT&T vice president, was perfectly positioned to introduce sublicensing as a standard procedure to meet competition. He needed all of his powers of persuasion, however, for the parent Bell firm initially tried to restrict the practice.

At AT&T, men such as Thomas Doolittle and President F. P. Fish refused to relinquish their belief in an uncompromised competitive victory for the Bell system. They continued to maintain that Bell should always be able to furnish better service at lower prices than competitors. Holding to this older, imperious view, managers of the parent organization at first refused to permit non-Bell companies to use Bell equipment on their lines. Derided as a "dog in the manger" policy by independent promoters, this restriction provoked deep resentment from irate citizens and limited the practice of sublicensing.[89]

The policies of AT&T upper management ran directly counter to the experiences of men such as Hall. As the Southern Bell president had discovered, in many cases "a local company can furnish service at a profit—whereas a large company could not furnish such service without a loss."[90] Hall advocated use of compromise contracts because he understood this economic reality and saw compromise as the only way around an intractable situation in regions like the South, places that did not conform to the northern model of telephone development. By his own estimate, 65 percent of the country consisted of semiurban or rural territories "whose business is developed and carried on under less exacting conditions than in urban or suburban [territories]." Such places, because they did not need service "of as high a grade" as urban areas, could be effectively served by small local companies.[91]

Hall presented his case to his superior, Frederick P. Fish, making it clear that he advocated compromise for strictly economic reasons. Citing the case of Piedmont Telephone, a North Carolina independent, Hall wrote, "it is interesting to see that the company appears to have made a substantial profit under rates and conditions which would have made the business unprofitable had we operated it directly."[92] Piedmont served six small towns, had a total of 750 subscribers, and earned a modest revenue of $14.00 per station. Yet this business, undertaken at rates below Bell's, produced an eight month's profit of $2669.83 on an investment of $27,000.[93] Even under the terms of a sublicense contract and with an increase in the firm's depreciation and construction accounts, Piedmont could still earn an adequate 5 to 6 percent return. AT&T engineer Joseph Davis confirmed Hall's analysis a few weeks later. Both men concurred that Bell could never outcompete the rival; the only answer was sublicensing.[94] By relying on local capital and organization in marginal areas like North Carolina, Bell could retain a measure of influence over the territory and continue to expand its system.

President Fish was not persuaded by his vice president's detailed letter and continued to restrict use of the contracts to special cases.[95] Only in 1907, with the return of Theodore Vail to the helm of AT&T, did parent company policy begin to change. Hurt by declining revenues, Vail put a stop to futile attempts to outcompete challengers in every market.[96] Not satisfied with a divided telephone market, however, the AT&T executive sent a circular letter to all operating company presidents in 1908, advising them to pursue sublicensing more vigorously in unremunerative areas. He even recommended that if necessary they abandon Bell exchanges to secure cooperative contracts. Following Hall's lead, Vail recognized that control of quality was what really mattered, and thus authorized sublicensees to employ apparatus from competing manufacturers on connecting lines so long as it met Bell standards. The new president declared that it was necessary only that connecting companies operate in a manner "not to impair the quality of service furnished over joint lines."[97]

Cooperation, not competition, was the new order of the day, but it was cooperation on Bell's terms. Sublicensing, as Bell managers made clear over and over again, was a useful means "to control [the opposition] and have [it] operated to our benefit."[98] To this end, the contracts gave Bell crucial levers of power over connecting firms. Bell had final say over questions of technology, equipment design, and operations, allowing it to direct the independents in accord with system requirements. The agreements stipulated that the sublicensee:

> Shall provide the proper switchboard or other connections for direct communications on the lines of the first party [Bell] and the stations of its said exchange; and shall so construct, equip, maintain, and operate its exchanges as not to interfere with or impair the efficiency of service furnished through the lines of the first party connected as aforesaid.[99]

Only under these terms were independents allowed to interconnect with Bell lines. Bell thus did not need to fear that connecting carriers would lower the quality of its system. With terms such as these, Bell was able to continue pursuing its goal of a single, interconnected system firmly under its direction, but without the expense and political danger of formal consolidation.[100]

To be sure that sublicensing remained favorable to his company's interests, Theodore Vail reminded operating companies to make their agreements compact and limited to a small area, so that outside firms could not establish their own regional systems.[101] Vail's strategy stressed licensing case by case the connections needed to complete the Bell system and joining independents with Bell operating companies and independents with independents as the situation demanded. Sublicensees received in return for their concessions a pledge by Bell to engage in no competition within a five-mile radius of their exchange, providing them with a virtual monopoly.[102] But the monopoly was strictly local, for the contracts specified that all ownership and control of connecting toll lines remained in Bell hands.[103] When no Bell toll line was available, the contracts prohibited unapproved interconnection with other carriers.[104] Thus, while allowing local capital a necessary role, sublicensing still left Bell the undisputed market leader. Indeed, local resources, dependent on Bell guidance, actually became tools for Bell system growth, able to penetrate places where the national firm could not profitably go.

Bell followed the strategy of sublicensing in all regions where conditions resembled those of the South. In the intensely competitive Midwest, sublicensing was employed extensively (see table 6.3). Here too compromise tactics and a new approach to marketing the Bell system to skeptical customers paved the way for Bell's resurgence. Even economically advanced New England had pockets of resistance to Bell system expansion, areas where competitors operated effectively (see table 6.1). New England Bell too employed sublicense contracts to link these areas with its system. Indeed, the contracts proved useful anywhere competitors demonstrated that they could serve customers better than could Bell. By 1908, Hall's basic strategy was being applied nationally. In that year, Theodore Vail announced the concept of universal service—

Table 6.3. Comparative Sublicensed Exchanges, 1908

Company	Sublicensed Exchanges
Central Union	500
Colorado Tel.	38
Iowa Tel.	198
Michigan State	191
NET&T	303
NY & NJ Tel.	0
NY & PA Tel.	129
Pacific Tel.	228
SBT&T	257
Cumberland	189
Wisconsin	209
S.W. Bell	369

Source: AT&T box 1364, Sub–Licensee Statistics, 1908; AT&T box 1348, Sub–Licensee Statistics.

Sublicense Statistics, 1906

Company	Total Subscribers	Number of Sublicensee Telephones
Southern Bell	107,071	27,106
Central Union	178,087	58,044
NET&T	222,605	27,126

Source: AT&T box 2025, Toll Line Statistics, 1906.

basically an extension of Hall's prescriptions for the South to the rest of the nation.[105]

The South had served as a laboratory for developing new competitive policies for other territories. The years of battle with independent firms made SBT&T managers experts at handling competitive problems raised by Bell system expansion into marginal economies. The territory itself proved a congenial environment for innovation. It was run by a well-organized operating company, but was located away from the main competitive fray in the industry. Thus in the South the stakes were lower and the costs of failure were smaller than in more advanced regions.

The benefits of this experimentation began to show as the years went on. As late as 1912, independents still controlled 42 percent of the national market, and they were still well represented in the South and Midwest (see table 6.1). But their market share had declined from its high of 1907, and the nature of competition had changed. Division of market replaced direct engagement. During the first decade of competition, telephone firms grew wildly. But by 1907, the number of exchanges had dropped significantly, as competition began to level off (see table 6.4). The total number of Bell and non-Bell telephones continued to grow, but with natural monopoly conditions prevailing in *local* service, each town retained only one telephone firm. Bell and the independents had squared off, each to their own territory.

This change was in accord with the nature of telephone competition, as Bell and its competitors appealed to different markets. The great advance Hall had made was discovering a way to use this division to his firm's advantage, gaining for Southern Bell the richest territory in the South. By 1907 the independents had been pushed to the margins of the economy. They operated in less developed areas, built smaller exchanges, and served customers who made less intensive use of the telephone (see tables 6.4, 6.5). After 1907, market division accelerated, as Hall's strategy became accepted Bell policy. The independents continued to grow, but they were confined to less desirable places and grew at a slower rate than Bell. Nor were they able to create their own systems independent from Bell supervision. Instead, they were linked to the Bell system, informally—through the corporation's dominance of technology and long-distance service—and formally through the sublicensing contract.

While the new strategy helped put an end to competitive threats to Bell's system in areas like the South, it also placed new strains on the Bell's managerial structure. By the early twentieth century, the corporation faced several important organizational questions: What was the proper relationship between the various units of the Bell system—the parent firm, operating companies,

Table 6.4. Bell and Independent Telephone Systems, 1902–12

Region	Bell Ex.	Bell Tel/Ex.	Ind. Ex.	Ind. Tel/Ex.
New England				
1912	625	928	106	428
1907	643	570	89	491
1902	473	323	126	100
Midwest				
1912	991	1,164	2,010	674
1907	954	718	1,979	606
1902	679	472	2,832	162
Mid Atlantic				
1912	1,001	1,353	861	495
1907	938	936	749	500
1902	946	396	785	177
South				
1912	377	565	223	659
1907	319	404	154	681
1902	108	380	458	121
US				
1912	5,853	869	5,662	643
1907	5,418	578	5,195	574

Source: Bureau of the Census, *Telephones, 1912*, tables 24, 30.

Note: Regional figures are simple average. Exchange figures exclude firms reporting an annual income of less than $5,000. This omission means that the number of independent telephones per exchange is inflated for the South and Midwest, which had many small companies. Tel/Ex. = telephones per exchange. See table 6.1 for regional definitions.

and sublicensees? Would too much decentralization jeopardize the goal of integrated service? E. J. Hall turned his talents to these important organizational matters as well during his tenure in the South.

Even before coming to the South, Hall had pondered the proper relationship between the parent and operating companies. In an 1886 letter he had expressed his fear that centralization and

Table 6.5. Comparative Telephone Use, 1912, Nonrural

Region	Bell Talk/Tel	Ind. Talk /Tel
New England	1,330	705
Mid Atlantic	1,294	1,556
Midwest	1,898	1,420
South	2,533	993
US	1,795	1,263

Source: Bureau of the Census, *Telephones, 1912*, table 25.

Note: Talk/Tel = messages per telephone per year (325 days). See table 6.1 for regional definitions.

compete control of the industry by AT&T might alienate local interests and result in a public backlash against the company. Rapid, unilateral organization from the top down could exact a heavy political price.[106] While Hall believed that AT&T should buy majority ownership in operating companies like Southern Bell, he did not want to abandon the territorial structure altogether. Instead, he advised AT&T to proceeded cautiously with reorganization.

In part this astute analysis allowed Hall to see that, in the South after 1894, proposals for radical reorganization such as those of Thomas Doolittle would not work. Like the AT&T engineer, Hall believed that it was "the duty of each company to study the relation between its system to the whole system." But Hall preferred to enforce this policy in more subtle ways. At Southern Bell, he forged new financial ties between parent and operating company to achieve coordination, as we have seen.[107] To complement this link, he also began to train and install men for key middle- and lower-management positions who were loyal to AT&T's plans. Men who had either been born in the South or had spent their entire career at Southern Bell assumed positions of responsibility under Hall. In 1883, the directors of the southern firm had been drawn mainly from Western Union and ABT. By 1901, E. J. Hall himself was the only outside executive. The others, D. I. Carson, James Easterlin, and W. T. Gentry, were either southerners or men with long tenure at SBT&T.[108] This personnel policy accorded with Hall's perception of the political dangers of outside control. As Bell moved aggressively to retake the South, southern managers proved very adept at quelling discontent from southern customers and parrying rhetorical thrusts from competing entrepreneurs.

Hall's approach to management and organization flowed from his entrepreneurial strategy. His opposition to greater central control reflected his understanding of the importance of regional variations in any large-scale system. Differences in supply and demand conditions, business climate, and political culture were simply an inevitable fact of the telephone business.

In addressing the border problem—the issue which had so plagued Thomas Doolittle—Hall made this point clear. Doolittle had maintained that Bell's territorial divisions inhibited necessary toll line construction. The AT&T engineer had proposed radical reorganization in response. Hall, to the contrary, held that long line development was perfectly in accord with "traffic demand and available capital," in each region.[109] In a letter to AT&T president Fish, Hall, using Southern Bell as his example, wrote:

> An examination of the toll line maps of a particular territory will show . . . development from commercial centers in radial lines with enclosed gaps between the groups, bridged partially by main lines bearing the same relation to the groups that the lines of our own company [AT&T] bear to the local system.[110]

Such a situation, according to Hall, indicated that existing toll line construction was economic and rational. Each company, "primarily interested in developing its own territory so as to secure the largest possible return will naturally and inevitably radiate from and feed towards its own business centers."[111] This very fact had seemed to Doolittle a weakness and had caused him to advocate dramatically changing parent-operating company relations. But for Hall it was an indication of the strength of Bell's existing structure.

The Southern Bell president had not ignored the need for central control. He maintained that some "central organization" was necessary "to adjust relations between companies and bind the various units into a comprehensive national system."[112] But AT&T could provide the needed glue to hold the system together simply by controlling the flow of capital to operating companies, as it did with Southern Bell. The parent organization also held key patents on supplementary telephone equipment and had, a decade or so earlier, formed an organization to control and coordinate technology. With these controls in place, regional firms themselves could be expected to carry out necessary development efforts. Some border problems might arise, but Hall opined that these could best be dealt with on a case-by-case basis. Overall, he felt, "the magnitude of the problem is more apparent than real."[113]

Hall recognized that alternatives to his recommendations were possible. As he noted, "it is undoubtedly true that if the entire United States were operated by a single company, many of the gaps which exist between the present local territories would be closed."[114] Why not then operate the telephone industry in such a fashion, replacing local with top-down control? First there were the political dangers Hall had elucidated twenty years earlier. Far from disappearing, they had grown greater under competition. Then too were the internal costs of such organization. Highly centralized bureaucracies can become inefficient. Few people at AT&T seemed aware of the costs of overcentralization, but I believe Hall had already perceived the danger. Engineers at AT&T tended to be very confident that they could organize and administer a fully centralized system without incurring significant costs and inefficiencies.[115] Hall was more skeptical. He preferred to rely on strategic levers of control offered by AT&T's command of capital and technology, as well as the presence of loyal, well-trained local managers. These controls he had installed at Southern Bell, and they were apparently working.

A flexible, partially decentralized structure provided the best means of assuring the quality of an integrated system in the face of very real and persistent variations in conditions across regions. This point went back to Hall's basic entrepreneurial strategy. The "real cause" of the existing pattern of the toll network, he made clear, was "inherent in commercial conditions and our general scheme of organization."[116] The scheme of organization referred to Bell's territorial divisions, which could not be changed for reasons noted above. More important was the point Hall made in the first part of his sentence. In many places like the South, Hall was indicating, demand was simply not sufficient to support more toll lines. If AT&T wanted to continue to build its system in accord with its high-quality standards—and Hall was clearly in favor of this policy—then it would have to accept these demand-side limitations.[117] The best means of dealing with them were those he had employed in the South—selective expansion of the network into commercial centers and use of compromise contracts in areas which continued to choose independent alternatives. The best structure for administering this type of system was a decentralized, territorial one.

By rejecting the rigid technological and entrepreneurial style which Bell had developed during the monopoly years, Hall was able to achieve in the South two desirable, but seemingly incompatible goals. He furthered the growth of Bell's integrated network while penetrating remote areas under the control of rivals. Engineers such as Doolittle had initiated the first phase of this development by producing plans and measures to operate the business on a systemwide basis, by-passing the artificial boundaries of geography. But geography, regional culture, and economic conditions did matter, as Hall discovered. Abandoning the old assumption that Bell's technical superiority gave it an unbeatable weapon, Hall produced important innovations. Cooperating with local businessmen in numerous small southern towns, he devised a complex marketing strategy that "sold" the Bell system to skeptical customers. Expanding the network selectively, winning back territory, carefully using compromise contracts, he satisfied southerners' desire for some local control without sacrificing Bell system standards.

Like his predecessor in the South, James Ormes, Hall also built an organization that conformed to his strategic vision. Ormes had created the first regional telephone operating company in response to southern backwardness and entrepreneurial apathy. Hall introduced a more decentralized structure to deal with demand-side constraints on system expansion. Relying on principles of cost accounting, functional organization, and decentralized administration, he reshaped the southern telephone industry in accord with the needs of the emerging Bell system.

As this chapter suggests, the formation of a single national telephone system under Bell control was not the product of the inevitable workings of technological forces or economic laws, but of human will and innovation. Other paths were possible. Local companies in the South and Midwest could have kept large blocks of territory beyond Bell system reach, cooperating to form an alternative system to AT&T's. As late as 1907 independents carried 20 percent of all toll calls, mainly in the midsized city market.[118] In addition, several large financiers from the East almost extended capital to the independents to help them wrest control of major cities from Bell as well.

Bell met each of these challenges. In the South and Midwest, the corporation depended on the policies pioneered by E. J. Hall to retain strategic control. In Chicago and New York, it blocked entry of competitors through its political influence with tunnel authorities.[119] When the "traction kings" led by financier Peter Widener tried to form a competing long-distance company, Bell resorted to its powerful friend J. P. Morgan to turn them back.[120] The company's overwhelming power in the telephone market—a product largely of its seventeen-year patent monopoly—combined with its innovative energy closed off other options and set the telephone industry on the path to a national communications network under AT&T control.

The success of AT&T policy offered certain benefits to the South and the rest of the nation. The corporation kept the industry united under a single standard of performance, and over time the separate territories of the telephone industry began to converge. In the nation telephone distribution increased threefold between 1902 and 1912 (see table 6.6).[121] Small, strictly local independents declined, and the competitors that remained became connected to the Bell system. Critics of the Bell corporation argued, of course, that these changes meant a slowdown in the rate of telephone growth and higher prices. But the remodeled industry was also more stable, better able to provide high-quality service, and more unified. These factors enabled Bell to help bring remote places like the South closer to the center of the economy.[122]

Yet the process of convergence was still incomplete, as the South remained the most backward part of the industry. Nor was everyone satisfied with Bell's reign, even if it was somewhat beneficent. Local entrepreneurs chafed in their subordinate role in the Bell system. Various parties in the South pondered other options for telephony. To consumers of independent telephone service who

Table 6.6. Comparative Telephone Use and Distribution, 1902–12, All Systems

Region	Telephones/1000	Talk/Tel
New England		
1912	93	1,293
1907	68	1,391
1902	27	1,689
Mid Atlantic		
1912	90	1,321
1907	73	1,556
1902	29	1,415
Midwest		
1912	135	1,639
1907	109	1,621
1902	50	2,196
South		
1912	32	1,925
1907	22	2,481
1902	11	2,698
US		
1912	91	1,573
1907	72	1,690
1902	30	2,138

Source: Bureau of the Census, *Telephones, 1912*, tables 5, 20.

Note: Tel/1,000 = telephones per 1,000 people; Talk/Tel = messages per telephone per year. See table 6.1 for regional definitions. Regional totals are simple average for 1902, weighted average for 1907, 1912.

did not need AT&T's long-distance lines, a divided telephone market seemed a preferable alternative, with Bell serving the more advanced regions and independents serving the sectors that valued mainly local connections. Interconnection between the two halves of the market, to the extent it was even necessary in places like the South, could have come about gradually, on more equitable terms.

There was no overriding reason why Bell system standards had to be the ones adopted for the industry. Conflicts thus continued between regional demands and system needs. Dealing with these remaining problems required another set of innovations. Hall had been successful in adapting the Bell system to the southern environment. He, and other men, now began to search for a means of fully integrating the South and similar regions into the Bell system in a way which eliminated this discontent.

CHAPTER 7

A Merging of Interests

AS SOUTHERN BELL approached the end of its first decade in the twentieth century, it appeared to have a secure position. During the last years of Hall's presidency, the southern telephone industry had experienced impressive growth. Between 1902 and 1907 wire mileage in the region increased 166 percent, slightly more than the national average of 165 percent.[1] For the South—usually so far behind the rest of the nation—such figures indicate a startling achievement. So too do the advances in crucial urban markets. Over the same period the percentage of telephones in southern cities had risen to levels comparable to those of other regions. The percentage of urban population with telephones was 7.8 in the South, 7.1 in the Midwest, and 4.9 in New England.[2] Hall's policies had assured that Southern Bell would benefit from this growth. Much of the increase in wire mileage, for example, was due to the construction of toll lines in the region. Even where Southern Bell did not build directly, it could be sure that competition would redound to its benefit in the long run, as more companies were acquired or made sublicensees of the Bell system.

Nonetheless, there were limits to the benefits the new policies could bring to Southern Bell. In particular, as the company moved out of major urban areas, system growth began to present new problems. Any further extension of the Bell network in the South necessarily involved significant forays into rural and hinterland markets. Only 361 of 25,000 southern towns and cities contained more than 1,000 residents. Rural counties and small communities clearly held most of the future southern telephone demand. Yet SBT&T had done poorly in these places, considerably worse than its competitors. So much of this underdeveloped territory was still in the hands of independents that Bell seemed likely to loose a significant percentage of its future profits as demand for service increased. Large segments of the South also remained outside of the long-distance network. If competitors in these areas built their own toll lines, they would slip out of Bell control altogether.[3]

Sublicensing and the compromise policies pioneered by Hall had staved off some of these dangers thus far, but their efficacy for the future was in doubt. Sublicensees would still capture profits from expanded service into hinterland markets. Decentralization, while a necessary response to varying economic conditions, could, if carried too far, undermine Bell's goal of creating a unified network. To maintain its market share and continue to build its network, Southern Bell had to integrate outlying areas into its long-distance system and do a better job of reaching out to farmers and rural residents.[4]

Of these challenges, rural telephony presented the greatest immediate difficulties, in the South and indeed throughout the nation. Rural service was relatively untouched by Hall's reforms. Noncommercial telephone firms, cooperatives, and independent farmer lines still predominated in the rural market, and they had done exceedingly well in serving their clientele.[5] As one Bell observer noted, in rural areas "the telephone investment is now being made by local capital." The key was, "to make this investment reinforce Bell investment." The time had come "to take the initiative in rural development."[6] This would be no easy matter. To reach into rural areas, Bell operating companies would probably have to vary their service standards, equipment quality, and rates to meet farmers' special demands.[7]

These changes ran counter to AT&T policy, which continued to stress a high standard of service quality. When competition had first arisen, AT&T personnel had advised against technological adjustments for small exchanges and outlying areas, fearing that they would harm the toll network.[8] Over the years, this attitude had changed very little. Even in the face of competitors' successful rural telephone systems, Bell agent W. S. Allen believed that "the small villages and scattered farmers do not want telephone service."[9] The conventional wisdom still held that Bell would do better maintaining its standards and concentrating its efforts solely on towns and cities. Thus, while AT&T managers feared losing to competitors vast stretches of territory in rural and semirural regions, they were still reluctant to adjust their organization and technology to serve these markets.

In 1904, AT&T President Frederick Fish made some small changes in company policy to help Bell firms expand into rural areas. Correctly noting that Bell prices militated against rural service, he suggested reducing rates on farmers' lines.[10] The chief executive did not back up his advice with any significant action, however, and limited his assistance to lowering the charge to operating companies for Western Union equipment, encouraging them to pass the savings on to farmers.[11] Bell operating companies, with their huge investment in lines and equipment, could never undercut small independent firms and farmer cooperatives in this way. The attempt would only damage their rates of return. Fish capitulated to the advice of AT&T managers, concluding weakly that "[rural service] depends so much on local conditions and the local atmosphere that its should be left to the ingenuity of the management of each company to adopt schemes for getting the farmers."[12] Unwilling to make more substantial changes in parent company policy, Fish abandoned the regional companies to their own devices.

Lacking strong leadership from above, Bell rural service stagnated in the early years of the twentieth century. Individual operating companies proceeded on their own into rural areas, with mixed results. Several firms, including Southern Bell, used special sublicense agreements with individual farmers and rural companies. Some members of the parent firm thought these arrangements unwise, however, because they ceded too much control of the system to local operations.[13] Caught between the farmer's demands for special types of service and AT&T's inflexibility, Bell rural telephony could make little headway. What was needed was a set of innovations for the rural market such as those pioneered by Hall for urban areas.

By 1906, Bell began to move in this direction, as the parent organization loosened its strictures on technology and operations.[14] Taking account of rural needs, AT&T engineers produced a standard set of equipment and procedures for rural lines. This new technology embodied principles of low cost and simplicity of use and maintenance. Maintenance was an especially great concern in places like the South, where the costs of servicing remote locations could be high. To keep these costs down, AT&T lowered the quality of its rural technology somewhat by employing earlier, simpler types of telephone equipment. Still fearful that such equipment would interfere with the rest of the network, the company built into its new rural technology special features to minimize interference with the rest of the system.[15] Rural offices employed small capacity switchboards and older magneto systems, which were both less expensive than common battery systems and not likely to interfere with other transmissions.[16] Making rural equipment too separate from the rest of the system, however, would have decreased overall efficiency. AT&T therefore designed its rural technology to hook into its urban exchanges.[17] In places where demand so warranted, a special rural operator handled incoming calls from farmers, but, to keep up switching efficiency also took regular calls from city subscribers when not otherwise engaged.[18]

Years of competition, public outcry, and threatened legislation had forced Bell to provide the special service that rural customers demanded. To do so the company had to vary its technology and lower some of its service standards; but as Western Electric engineers discovered, many of the earlier fears of such changes were based on shortsightedness and prejudice. It was possible to meet farmers' demands and still build in enough safeguards to protect the integrity of the network. The result was a compromise between Bell engineering standards and rural residents' desire for low-priced service.[19] By adjusting its technological style in this fashion, AT&T enabled operating companies such as Southern Bell to enter the rural market.

Using the new technology, Southern Bell began to devise plans for attracting rural customers. In 1909 Southern Bell president W. T. Gentry declared that his firm was making every effort to furnish service "adapted thoroughly to the needs of rural communities but stripped of every element . . . non essential to the country [which] would tend to increase the cost beyond a point at which the farmers could afford to take service."[20] The company got urban businessmen to build rural lines from their cities to nearby farmers, thereby lowering the cost of rural expansion by relying on local capital.[21] Other Bell companies developed similar methods to employ local capital and entrepreneurship to extend their systems into rural counties.[22] While Gentry also advised farmers to form cooperatives and build their own lines, he warned that such service "is by no means standard telephone construction and would not be adapted for use on toll lines nor urban service."[23] Clearly Bell would maintain its standards. Nonetheless, by reforming its rural operations Southern Bell managed to triple its number of rural stations between 1907 and 1909.[24]

While these changes in policy improved Southern Bell's performance in rural areas, they alone were not sufficient to extend the Bell system into other markets still dominated by independent firms. This had been a prime goal of E. J. Hall, who had hoped that by reforming operations he could expand the Bell system into a wider variety of places. After 1907, however, achievement of this goal appeared increasingly problematic. Between 1907 and 1912, Southern Bell continued to push into the southern hinterland, but expansion slowed. Measured by increases in wire mileage, the southern network grew by only 28 percent in these years, compared to the national average of 56 percent.[25] Earlier, when there had been much competition in the industry, growth had been rapid, as Southern Bell and its rivals fought for customers. By 1907, Southern Bell had substantially increased the size of its market through acquisitions and sublicensing, but also reduced competition and slowed down expansion. To some, this fall-off in activity indicated that the corporation was again becoming a complacent monopoly. But the nature of expansion had also changed. Having taken the most important urban markets and closely related rural ones, Southern Bell was approaching areas of truly marginal demand for its system. Future telephone growth in the South would involve less dramatic increases; yet there were still important benefits that the Bell system could offer the South. If Southern Bell could consolidate its various markets and its different types of service, it could provide unified service that linked the South more closely to the rest of the nation.

This type of system building required a new approach to toll line construction. As Southern Bell auditor J. B. Hoxsey noted in 1908, a large percentage of southern toll calls traveled less than fifteen miles.[26] Perversely, however, the southern network consisted of many long trunk lines and few short feeders. In many places, moreover, toll and local service remained separate. Hoxsey recommended that Southern Bell build and connect more exchanges within the fifteen-mile radius to round out its network. Only then would the southern system produce the returns of which it was capable.

This new construction was also part of a larger plan to build demand for long-distance service, something both AT&T and SBT&T greatly desired. Southern Bell's toll network was relatively strong in urban markets, where customers could afford the high rates it required. But beyond these areas, in the South's economically underdeveloped hinterland, long-distance demand remained weak. In such places the firm relied extensively on its sublicensees, as well as non-Bell carriers, to provide service. The challenge was to bring these areas, with their large number of existing and potential customers, into the Bell network.

Hoxsey believed that here too his plan could help. By building short lines out from urban centers to the countryside, Southern Bell could gradually increase its toll business. In such areas it would find a small but eager clientele for this service. By making many short connections, the company would be in a position to draw toll business from its own outlying exchanges and those of its many sublicensees and interconnected non-Bell firms. They could supply the demand needed to support the new lines. Drawing on the compromises enacted in previous years and the recent growth in Bell rural telephony, this strategy would increase southern toll revenues and create a more fully integrated system.

The key to this plan was keeping the price of toll service low in order to encourage long-distance communications. Hoxsey recommended reducing rates from twenty-five to fifteen cents on short-haul calls, believing demand to be elastic enough to generate an overall rise in toll revenues.[27] In addition to increasing demand for toll service, this new price structure would reduce public outcry over Bell rates.[28] Though a necessary change, the decrease in charges could not be allowed to damage SBT&T's overall rate of return. To keep profits up, therefore, Hoxsey suggested ways of inexpensively increasing the speed, efficiency, and capacity of the toll network.[29] He advised Southern Bell to use phantom circuits to augment line capacity quickly and cheaply, as had the Wisconsin and the Kansas and Missouri Bell companies.[30] Rapid construction of inexpensive circuits which could be easily upgraded when usage increased in the future, would also help to expand capacity without unduly burdening Southern Bell's revenues. These policies too had worked well in areas such as Colorado, Oklahoma, and the Rocky Mountain territory. All of Hoxsey's recommendations were aimed at one thing—economically lowering the price of toll service in order to expand the scope of the southern system.

Though the plan seemed workable, Southern Bell moved slowly to implement it. Six months after Hoxsey's report, E. J. Hall remained uneasy about the progress of the southern network.[31] Despite Hoxsey's thorough analysis, rates remained a problem.[32] For his plan to succeed, Bell would have to reduce the price of toll calls, but keep up the price of local service. Exchange rates had dropped precipitously during the years of fierce competition, and municipal authorities, sorry to see the return of monopoly conditions, were only gradually letting them rise again.[33] With local rates unstable, reductions in the price of toll calls could hurt Southern Bell profits. Under these conditions, the firm could not aggressively pursue toll line construction. Eventually Southern Bell, and other companies, relied on regulatory authorities to steady prices.[34] In the meantime, however, expansion proceeded slowly.

To spur system growth in these years, AT&T transferred certain toll circuits it operated to Southern Bell. The regional company could run them more cheaply than the parent organization. In addition, small southern stations, which by themselves were not profitable to maintain, would pay if tied into these kinds of toll lines. Income from the toll circuits would also stimulate Southern Bell to build more feeder lines to AT&T trunks.[35] For similar reasons, Thomas Cotton also advised Southern Bell to purchase certain independent firms and sublicensees whose short-haul lines paralleled Bell's own.[36] Fifty-seven in all, they made considerable profit on the short-distance interexchange business. If Southern Bell enjoyed these revenues, it would have further incentive to build long-distance lines.

Cotton believed that Southern Bell had to take greater control of non-Bell firms' operations if it was to continue its expansion into the vast, untapped hinterland.[37] Some stations in the hands of independents were located at key long-distance switching points. It was imperative, Cotton maintained, that Southern Bell itself perform this work, "in order to properly handle, supervise, and check the toll business."[38] In Georgia and Alabama, the hubs of the entire southern long-distance network, sublicensees and independent exchanges outnumbered Bell three to one.[39] Non-Bell firms in these states had built a considerable toll network of their own, and made substantial profits handling crucial short-haul traffic. Cotton feared, with good reason, that if they were allowed to continue in this fashion, they would break lose from Bell control altogether. Key acquisitions would decrease this danger and assure that "the development of sub-licensee systems will be a source of strength to us."[40]

Southern Bell could not, of course, acquire all non-Bell firms. Some of them operated in territories it could not profitably serve, as Hall had demonstrated earlier. Even while making strategic purchases, therefore, Southern Bell continued to add sublicensees to its organization. Between June and December 1909, it signed forty contracts with independent firms in Georgia, Alabama, and North Carolina.[41] Purchasing some firms, making compromises with others, and in certain circumstances releasing its own exchanges to still others, Southern Bell worked toward the proper organizational structure for the southern market. At the same time, by constructing short-haul toll lines, further connecting sublicensees and outside firms to its circuits, and pioneering new policies on rural telephony, the company penetrated remote areas. The result of this diverse set of policies was, as Thomas Cotton put it, to "rationalize" the Bell system in the South by bringing together local systems built to serve specific local markets with AT&T's national network.[42]

While crucial to overall Bell strategy, these new policies

raised additional problems. New rural technology, the release of circuits from AT&T to Southern Bell, more sublicensees, increased acquisitions, though all necessary parts of expansion, threatened to undermine central control of the Bell system. Engineer Hammond Hayes no doubt spoke for many at the parent firm when he stated that rural operations "wasted" labor and increased costs.[43] For the engineer, who wanted to plan the system strictly in accordance with the principles of engineering efficiency, rural systems and other adjustments of operations to serve marginal areas were indeed a waste. Yet while these policies seemed inefficient and costly, they were a necessary element in Bell system growth.

Like a recurring malady, the conflict between central control and local responsiveness again plagued Southern Bell. During E. J. Hall's years in the South, this problem had been confined to the regional level, where Southern Bell had found that in many cases it was unable to compete with independent firms because of its inflexible commitment to a single standard of service. Hall had resolved this issue by decentralizing operations through contracts with local firms. After 1907, however, the conflict moved up a level, pitting SBT&T and AT&T against each other. Southern Bell itself had successfully built a structure that combined local flexibility with central control at the regional level. But as the firm continued to link outlying areas with the network, it created a more distended Bell system in the South, prompting criticism from AT&T management.

With further expansion in the South taking place in even more remote and marginal areas, the split between Southern Bell and AT&T was likely to widen. Operations in these areas would require further modifications of Bell standards and more decentralization. Nor was the problem limited to the South. Throughout the nation, regional Bell firms made different decisions about technology, methods of construction, and operations as they responded to conditions in their markets. If this situation continued, it could easily subvert AT&T's goal of creating a unified network. To combat this danger, parent company managers in 1907 began to reform the structure of the entire Bell organization, seeking new methods of central control. This new structure, influenced by E. J. Hall's earlier work in the South, reset the balance between operating companies' need for flexibility and AT&T's quest for standardization.

Change began in 1907 when AT&T itself underwent a major reorganization. In that year, Theodore Vail returned to the presidency of the company at the behest of J. P. Morgan. Over the last years of the Fish administration, Morgan and several other financiers had gained control of AT&T.[44] Hurt by falling profits, the new owners had formed in 1906 a committee of Vail, John Waterbury, George Baker, and several others to consider what could be done about righting their listing enterprise.[45] Though the deliberations of this committee are not known, not long after it met Vail replaced Fish as president and instituted a three-column functional plan at AT&T.[46]

The functional scheme offered AT&T better means of control over its sprawling enterprise. Reorganization began in the Long Lines department of the company, where ten departments and seven operating units were replaced by three specialized line units—plant, traffic, and commercial—coordinated by a central staff. These new, functionally arranged departments were then subdivided into regional and local divisions, based on broad new territories.[47] The new structure made lines of authority between departments clearer and information necessary for strategic planning more accessible. As AT&T managers explained, functional organization permitted economies of specialization and allowed coordinated development.[48] With Long Lines under functional control, AT&T applied the same structure to its manufacturing arm, Western Electric.[49]

For Southern Bell, these developments had several important ramifications. First, Vail began to provide the top-level leadership so noticeably absent during the Fish administration. The new chief executive quickly endorsed J. B. Hoxsey's plan for expansion, believing that it addressed fundamental problems of growth at the margins of the Bell system.[50] Though AT&T was short of funds, Vail also found money to aid southern development. The parent firm lent Southern Bell $2.6 million in 1907.[51] In early 1908 it extended an additional $1.65 million for construction and reconstruction, as well as $1.38 million in short-term credit.[52]

Southern Bell itself adopted the functional structure in

1909, as the new form of organization spread throughout the Bell system.[53] The plan aided the southern operating company in controlling its extended operations, just as it assisted AT&T in the same task. Southern Bell, however, made several important modifications of the new scheme. Some AT&T managers, taken with functional organization, believed it should be applied down the line, eliminating territorial divisions altogether. As they saw it, over time the same functional forms could be replicated at the level of the local exchange, enabling Bell to run the telephone business as a single, centralized organization.[54] These prescriptions proved to be overly ambitious, however, and Southern Bell combined the advantages of functional organization and centralized operations with the important flexibility it had achieved earlier through decentralization.

In forming this combined structure, Southern Bell drew on the experiences of other operating companies in the Bell system. These firms carried out the invaluable work of testing functional organization under a variety of economic conditions. Bell firms in urban and industrial areas, for example, quickly adopted the new structure. In their territory, long-distance demand was growing rapidly—prompting the need for a new method of operations—while markets were fairly homogeneous, allowing for the efficient application of the new structure. Firms in the most advanced sections of the nation—New York, New England, the Pacific Coast—conformed closely to the AT&T model.[55] Indeed, in 1899 New York Telephone had tested the new structure before it was put into place at AT&T.[56] New England Bell moved steadily toward full functional organization between 1899 and 1908.[57] In 1902 the firm added a general superintendent and a chief engineer, replacing divisional superintendents; in 1906 it created the office of general superintendent of traffic and placed the engineering department one notch closer to the general manager's office. Finally, in 1908 New England Bell had a three-column functional form.[58]

Elsewhere application of the new principles of management proceeded more slowly. By 1910, Wisconsin Telephone had only a partially functional structure.[59] In Colorado, E. B. Fields protested that functional organization would not work in "thinly settled regions" such as his.[60] Fields noted that specialization in such places carried certain disadvantages. In his remote exchanges, he needed men with a general knowledge of the telephone business. Bureaucratization, while effective in advanced areas, would, he maintained, generate public resentment against a cold and impersonal management in Colorado.

Southern Bell president E. J. Hall provided the most sophisticated analysis of the limits of functional organization. He remained chief executive of this company until 1908, while continuing as AT&T vice president. In the latter role he was intimately involved in the changes taking place at the parent firm. Drawing on important information coming in from the field, as well as his experiences in the South, Hall stated bluntly that AT&T must not follow "the assumption, which I consider wholly unsound, that the method of organization can be successfully expanded indefinitely."[61] The vice president had long understood that, while territorial divisions might be inefficient for central planning and control, they could not be eliminated without significant costs.[62] A fully centralized company was unwieldy and incapable of responding to the different conditions of individual markets. Like Fields of Colorado, Hall recognized that specialization had its limits. Organization, in his view, was not a matter of a priori theory, but a means to an end worked out through experience. "Existing machinery," he had written earlier, had to be "adapted to existing conditions."[63]

Based on Hall's keen insights and the experiences of other firms, Southern Bell formed its new structure. Though organized functionally, the company continued to rely on sublicensing and compromises with competing firms. Some of these companies were acquired, but most remained financially separate from Southern Bell. Most were also not organized functionally, but continued to operate as small, local concerns. Southern Bell thus took advantage of the greater central control that functional organization allowed, but preserved needed responsiveness to territorial conditions and local needs.

The actions of firms such as Southern Bell demonstrated that regional conditions still affected telephone industry organization. Expansion of the Bell system would not eradicate all differences between regions, nor eliminate all vestiges of local culture. Advantages in specialization and greater central control had to be balanced against the costs of bureaucratization. In some markets, the costs outweighed the benefits. Some operating companies served markets too small to take advantage of the new principles. In others, organizational inertia blocked change. In still others, competitive threats made local managers reluctant to embrace new ideas.[64] While many of those who at first had opposed reorganization eventually went along with the new plan, Fields and Hall had correctly assessed its limitations. Only under certain economic conditions did functional organization increase efficiency. Profitability, not abstract organizational ideas, had to serve as the guide for applying the scheme throughout the Bell system, and the system itself had to adapt its technology, operations, and structure to the various environments in which it operated.[65]

These same tenets held in AT&T's relationship with its own territorial companies.[66] As Hall expressed the issue to Theodore Vail in 1909, no matter how much Bell reorganized its operations, much of the business still had to be conducted "by a central organization through administrations which have developed along lines not laid down in any comprehensive plan."[67] Reorganization should involve not tearing down existing structures, but working through them. AT&T had to "find an effective method of administration through sub-company organizations, controlled by a general staff."[68] The limits of central organization required "at least in form, and also in substance," that the Bell structure "follow local organization lines."[69] Otherwise, the results would be "inefficient, wasteful, and disastrous." Hall did not deny that some greater central control over operating companies was still necessary to build an integrated system, but his caveats made clear that not one but several types of organization were necessary in the Bell system.

Following Hall's reasoning, Theodore Vail worked to combine central control with decentralized operations throughout the corporation, rejecting the advice of those who wanted complete centralization. The result was a particular policy which he termed "universal service." Though as the phrase implied, there was one system and one policy in telephone service, Vail left room for local responsiveness. Aware that Bell operating companies would resist total consolidation and the break up of their territorial units, Vail left parts of the business in the hands of regional managers. AT&T

set standards and guidelines for technology and operations, but let the actual operations of each territory conform to regional conditions. Vail also began to take into account other environmental factors such as politics and the law. To avoid antitrust entanglements and smooth over conflicts with the public, the AT&T president accepted state regulation of telephones. In all of these ways, he adapted the Bell system to the variety of regional conditions in which it operated.

The most important applications of these new principles came in the area of technology and engineering. In this department Vail brought together Bell's long preoccupation with central control of technical operations with its more recent realization that technology had to be allowed to vary somewhat across regions. The task proved difficult, for since the 1880s the parent organization had been seeking greater central control over operating company technology to prevent conflicts such as those that had beset Southern Bell. In the 1890s, AT&T had begun standardizing telephone equipment.[70] By the end of the decade, AT&T and Western Electric were cooperating to produce standard designs for switchboards and other apparatus for regional firms. Gradually a common professional ethos, created by bringing together engineers of parent and operating companies, pervaded the Bell system, building into it crucial lines of authority.[71] These lines now had to change.

The strains on these informal arrangements were clearly evident in the early twentieth century. While some Bell companies saw greater central control as a boon, others found it a burden.[72] Operating company engineers were caught between AT&T's concern with standards and their own firms' interest in local conditions.[73] Some companies wanted AT&T to handle all planning and engineering, believing that the parent organization could do it best.[74] But others objected, finding parent company designed equipment inadequate for their needs. Firms like Southern Bell, which had many small exchanges, discovered that it was very difficult to set standards for exchanges of less than 2,000 subscribers.[75] In such cases, general specifications rather than detailed plans had to do.[76] Slowly AT&T engineers began to realize that not all territories could use the latest technology. Yet they still sought some way to at least make equipment compatible and comparable across firms.[77]

Between 1900 and 1905, therefore, AT&T reorganized its technical departments in order to maintain the levers of central control it had established earlier, while allowing greater local company input into equipment design. The corporation simplified engineering arrangements, bringing its own technical departments and Western Electric's together under one roof. AT&T also adopted a long-term perspective on equipment and apparatus.[78] Through new planning methods, the corporation's engineers were able to calculate the minimum cost of an exchange over a fifteen to twenty year horizon. This long-term perspective allowed operating companies to build systems which over the long run conformed to Bell standards, despite short-term fluctuations in demand.[79] By this means, AT&T helped to assure that all company equipment and operations conformed to system needs. As AT&T engineer Davis explained, however, only certain types of standards were necessary to prevent interference along the system.[80] Parent company engineers were only to provide models and general plans; the actual details of operations were left to individual operating companies.[81]

In 1907 Theodore Vail reorganized AT&T engineering once more, picking up the two strands of policy—central control and local responsiveness—and weaving them into a tighter pattern. The parent organization continued to use Western Electric engineers to assure that equipment in the field fit Bell system needs.[82] AT&T engineers also gained greater authority over inspection and the power to set systemwide standards for operating company equipment.[83] Yet while top-level control was tight and lines of authority clear, daily operations were left to operating companies themselves. AT&T proposed guidelines and models, reviewed work, made and received suggestions for improvements, and took a strategic, long-range view of planning. But this approach still left plenty of room for local flexibility.[84]

Financial policy was a second important area where Vail reshaped company policy. Changes in finance, however, followed a somewhat different course than that of technology, involving greater increases in central authority over time. In this area too, however, some local control remained. Between 1899 and 1906, AT&T had obtained full or majority ownership of almost all its operating units. Through equity purchases and long- and short-term loans, it became the Bell system's major capital supplier.[85] Between 1900 and 1903, AT&T also standardized its system of accounts, providing its staff with more information on regional companies' operations. In this way, the parent organization was able to compare and evaluate the performance of individual units, using profitability as its basis for strategic decisions.[86] During the years of rapid expansion following 1900, these financial procedures helped AT&T to survive competition and begin consolidating the industry.

After 1907, Vail took the final step in this financial evolution. He stabilized the firm's financial position by increasing its equity capital, acquiring the necessary funds to continue consolidation. Instituting new methods of fiscal control, he also set stricter guidelines for investment and monitored more closely regional companies' construction budgets, eliminating items which were not generating sufficient returns. Although AT&T had final approval of operating companies' expenditures, some local control remained. AT&T monitored profit centers, but allowed regional company managers to operate semiautonomously in responding to local commercial conditions. Regional firms also continued to perform the necessary tasks of collecting information, filling out forms, and estimating their capital requirements.[87] At the same time, sublicensees and non-Bell firms drew on local resources in their operations, as AT&T found it unprofitable to incorporate these organizations fully into its system. In financial planning, as in other areas, the national firm relied on key levers of central control and decentralized operations.

With the functional structure in place and reorganization of technical and financial departments complete, AT&T executives felt confident in sanctioning practices such as sublicensing. Vail encouraged use of sublicense contracts by liberalizing restrictions on equipment placed on connecting lines, and by advising companies to employ local capital and cooperate with local entrepreneurs in unremunerative areas.[88] In these prescriptions, Vail essentially reiterated Hall's arguments of the previous years. Meanwhile, Hall himself adapted his views to the notion of universal service, stating that this policy depended on sublicensing and other forms of compromise to adjust the "universal" system to the different markets and demand conditions it faced.[89] Accordingly, the vice president advised greater parent company supervision of sublicensees, to be sure that these units too conformed to Bell policy and standards and did not threaten the "universality" of the system.

The result of all of these changes was a new consensus on firm structure and competitive strategy that helped to increase the number of companies connecting to the Bell system after 1908. By 1912, 63.5 percent of all independent telephones were connected to the Bell system.[90] Under pressure from the Justice Department, Bell in 1913 went a step further. It agreed in the Kingsbury Commitment to open access to its network to all telephone firms with which it was not in direct competition.[91] Non-Bell carriers were allowed to connect with the network so long as they used high-quality equipment, built metallic circuits, and looped into Bell trunk lines.[92] Nonetheless, the government-induced liberalization of interconnection fell comfortably within the lines of Vail's basic policy.

With the new controls in place, AT&T could afford to be generous, confident that competitors could no longer play an independent role in the telephone industry.[93] The compromises did not threaten Bell's basic dominance of the telephone industry because the government did not force the company to open access of its lines to independent long-distance companies. As Vail had made clear, Bell was going to be sure it retained "absolute control of the toll line business connecting exchanges."[94] Local companies, responding to local conditions, served as useful adjuncts to Bell firms, but did not challenge overall Bell control.[95]

By combining central control and organization with decentralized operations in these ways, AT&T management resolved many of the conflicts that had arisen in the telephone industry after 1894. At the top levels, the parent organization made key entrepreneurial decisions and undertook strategic planning. Free from daily worries, it could hopefully have a clear overall view of company needs.[96] Technical and financial controls linked the various parts of the Bell company to this central staff. Standardized procedures and practices and shared values instilled a sense of loyalty in all personnel to the greater designs of the Bell system. Yet operating company managers still retained enough autonomy to decide what was the most efficient type of organization and operation to serve local markets. In this way, the various units that made up the Bell telephone company merged into a single, flexible organization.

For Southern Bell, the new structure helped to close many of the gaps that had placed it in conflict with AT&T. SBT&T policy could now conform to southern commercial conditions, confirming the work Hall had begun. After Southern Bell adopted functional organization in 1909, it also lost its dependency on outside management. Hall resigned as president and W. T. Gentry, a career Southern Bell man and native southerner, took office. Under him, J. Epps Brown, also of the South, served as vice president. Though Hall and Vail continued as members ex officio of the executive board, the company was thoroughly in the hands of southerners or men with long tenure in the region. For the first time in its history, Southern Bell could train and develop its own managerial staff.

Yet important as these organizational reforms were, they did not resolve all of the problems connected with building the Bell system in the South. While Southern Bell was formally part of a national organization, the market it served still differed significantly from the rest of the nation. Overall, the South remained the most backward part of the industry. In the region, telephone use had increased significantly since 1902, but still remained well below the national average (see table 6.6). The average size of a southern exchange was still smaller than the rest of the industry, reflecting the region's lower distribution of instruments (see table 6.4). Within the South as well, disparities characterized the business. The region's many non-Bell firms served outlying areas, while Bell operated in central places. Rural telephony in the South, though improved, still lagged considerably behind urban service.

Decentralization had been instituted, of course, to deal with just these conditions. By dividing the industry into different markets, Southern Bell was able to take account of regional variations in demand. Decentralization combined with reorganization proved successful in overcoming much of the conflict between the need for technical and economic integration on the one hand, and the need to respond to the particular environmental conditions of different territories, on the other. But these policies also provoked political responses from southern consumers, entrepreneurs, and municipal authorities in less-developed sectors, who still saw no reason why their locality had to conform to Bell system standards. Even under the new flexible regime, there was no disguising the fact that the southern telephone industry was in the hands of a large, distant northern corporation.

The inability of the new policies to deal with these political responses proved to be their most significant shortcoming. With demand for telephone service different in different submarkets of the South, conflicts between system design and local needs remained hidden just beneath the surface. As Southern Bell continued to consolidate operations, it faced increasingly angry city officials, who demanded that service conform to the conditions of their particular locality. In response to these sorts of demands, Thomas Cotton had recommended that SBT&T acquire more sublicensed companies and independent firms.[97] Yet formal acquisition and consolidation only further angered city officials, who saw rates rise as Bell reestablished market control.[98] The return of Bell dominance, whatever it promised to bring, was also reinstating monopoly conditions in the southern telephone market, with the concomitant rise in prices and fall in output. By 1910, the disenchantment of city governments with these developments swelled into a threat to the Bell system, stretching the system's flexibility to the breaking point. In response Bell turned to other public authorities for the powers it needed to complete its system.

Headquarters of Southern Bell in the early twentieth century, Atlanta, Georgia.

Anderson, South Carolina, telephone exchange. Even in small southern towns, the high degree of skill, precision, and middle-class deportment required of Bell system operators is evident.

Central office plant, Burlington, North Carolina.

Thomas B. Doolittle. The long-time member of American Bell/AT&T had pioneering ideas about long-distance telephony, but he failed to appreciate the special problems of building a system in regions like the South.

Edward J. Hall. An innovator at both AT&T and Southern Bell, Hall established policies that made possible the growth of the Bell system in the South.

Cortlandt Street exchange in New York City, 1893. A high-volume urban exchange had technical requirements far different than most of those of the South, as can be seen below.

Central office, Columbus, Georgia, 1914.

Theodore Newton Vail, ca. 1915. After his return to AT&T in 1908, Vail reorganized the Bell system, incorporating many of the ideas and policies employed by Hall in the South.

Laying cable in Spartanburg, South Carolina. Underground cables began to replace crowded poles in the South in the early twentieth century, as they had in northeastern cities a decade or so earlier.

Southern Bell motor vehicle, Raleigh, North Carolina, 1903.

Southern Bell truck, ca. 1915–20. The advent of motor vehicles made service in rural regions like the South much easier.

Chapter 8

A Role for Regulation

Although AT&T had exerted a powerful influence on the course of telephone development in the early twentieth century, it was becoming clear by the early 1910s that private action alone could neither guarantee the company's place in the industry nor resolve all the controversial issues that competition had raised. Particularly important was the still-unresolved question of system design—would the nation have a single system of telephone service, and if so, what standards would that system follow? Though AT&T had gone a long way toward standardizing service, variations in economic conditions between regions still made this a vexing problem.

There were good reasons for favoring a single interconnected system. Rising incomes and expanding distribution of telephones would, Bell managers and others believed, prompt people to demand better service and more connections, even in places like the South. It was not clear, however, that the market alone could satisfy these demands. Even if consumers were willing to pay the costs of an interconnected network, the problem of externalities could block its completion. In general, the value of a system to all its patrons rises as more people are connected to that system. In deciding whether or not to take service, however, customers generally do not consider this factor, but instead base their choice on the value of service to themselves alone. As a result, they tend to underestimate the total social utility of a system. In places like the rural South, where customers had a strong dislike for the northern-owned Bell Company and where many isolated places still remained outside the scope of its system, this problem was especially acute.

At the same time, the prospect of an interconnected system under the control of one company raised disturbing visions of monopoly. Having enjoyed the fruits of competition, customers in regions like the South were reluctant to give them up for some abstract, long-run advantage. The advantage of the system was a debatable one, dependent on how users valued different levels and types of service. In the early twentieth century, many customers, having been subscribers to telephone service for only a short while, were still learning the possibilities of the technology. All of these conditions—the limitations of the market, fears of monopoly, the volatility of consumer demand—finally pushed America to a new type of public regulation of the telephone industry in the twentieth century.

The regulatory policies of this era differed in important ways from earlier public intervention in the telephone industry. Early on telephone companies were seen, in the popular mind if not in law, as common carriers that, like railroads, could affect a broad public interest. On this basis public officials in many places had for decades overseen telephony in a general way. Only with the start of competition, however, did they begin to shape the industry to a significant degree.[1] Between 1894 and 1920, municipal, state, and federal authorities established rules, standards, and guidelines for service. Each of these authorities approached regulation in a different way, employing a different regulatory philosophy; but all three addressed the main question in the industry—what type of telephone system would the United States have?

They answered this question largely in Bell's favor, setting to rest the controversial issue of telephone system quality. Other regulatory concerns, such as competition, rates, and profits, were of secondary importance.[2] Instead, regulators sympathetic to Bell's interests played a key role in homogenizing telephone service, bringing all areas up to the standards set by AT&T, and thereby helping to integrate the nation's still disparate telephone industry into one system. In this sense, public regulation of telephones was the last stage in the process of "selling" the concept of a national network to resistant places like the South.

Initially, not all public authorities supported the Bell system. The corporation at first had to fight against municipal governments which challenged its autonomy. While Bell continued to build its national network, local governments strove to make telephone service responsive to local welfare. Aligning with like-minded independent telephone promoters, who sometimes sat on city councils themselves, many fiercely independent towns and cities developed their own local telephone systems and held out against Bell encroachment.[3] Even when city governments sanctioned Bell service, they extracted compromises from the company that met what were seen as crucial local needs. In either case, Bell confronted public authorities hostile to its plans for building and operating a technologically integrated system. Much of the early history of telephone regulation involved battles between municipal officials and Bell managers over these issues.

Patterns of Bell-municipal conflict followed closely the shifting tides of telephone industry competition. In the South, where as we have seen competing firms arose rapidly to challenge Bell, independent-minded municipal governments quickly mustered considerable regulatory power and put it to use. Fearful of burdensome rates, they used their authority to extract rate concessions from telephone companies.[4] Montgomery, Alabama, supported W. F. Vandiver and F. M. Billings's independent telephone company against Bell after the promoters agreed to comply with an ordinance to keep yearly rates down to $24 for residential and $30 for business customers, and to provide the city with one free cross arm on its poles for police and fire lines.[5] Other cities in Alabama set similar regulations. Huntsville passed an ordinance fixing rates and specifying one free government telephone for each 100 installed, while Birmingham simply prescribed rates for service.[6] Lake City, Florida, took a different approach, sticking with Bell in exchange for concessions from the company. The city agreed to patronize Southern Bell's exchange exclusively, if the company charged prices no higher than those of comparable places.[7]

By keeping the price of municipal telephones low, authorities reduced public expenditures, a prime concern of growing New South cities. In 1899, Tampa, Florida, got Southern Bell to provide a discount of 25 percent on twelve municipal telephones, as well as a 33 percent discount on municipal telephones connected to metallic circuits, an expensive type of line.[8] Similar considerations induced Jacksonville, Florida, to serve its public treasury by passing a 1 percent tax on gross telephone receipts in the city.[9] Bell, however, fought these efforts. The firm had to support a growing, increasingly integrated network in which costs could not be easily divided and assigned to local systems. Rate regulation, reductions on certain classes of telephones, and taxes, the corporation's managers maintained, stifled system growth by lowering the firm's overall rate of return and tying telephone prices too closely to local costs.

Other conflicts arose in the South as cities and towns combined rate regulation with efforts to ensure that telephone technology and service conformed to local needs. While some municipalities, wanting toll connections, accepted Bell rates and service, disputes still arose over the nature of toll service. Small southern towns valued lines primarily to other nearby towns. Bell, however, was attempting to link these municipalities into its national network.[10] While Bell opened a long-distance trunk line between Atlanta and Washington, D.C., citizens of Anniston, Alabama, clamored for a connection between their town and nearby regional centers such as Atlanta and Birmingham.[11] Differences in perspective such as these led to frequent clashes between southern municipalities and Bell over issues of service and prices.

Even when southern towns and cities acceded to Bell system requirements, they kept local concerns at the forefront. Valdosta, Georgia, petitioned Southern Bell to build an exchange using advanced technology such as the common battery telephones, multiple switchboards, and metallic circuits. The city would not permit Bell to raise rates until the work was completed, however.[12] Carrollton, Georgia, also acted cautiously as it shaped local service. The city switched from an independent company to Bell, but maintained a firm hand in running the local telephone exchange and permitted rate increases only when matched by improvements in service.[13] Gibson, North Carolina, offered a five-year exclusive franchise to any company which would build a system using high-quality metallic circuits and provide long-distance connections but charge no more than $2.50 per month, prices below

Southern Bell's normal monthly rate of $3.00 to $5.00.[14]

Larger southern cities presented even stronger challenges to Bell. While the company generally outcompeted independents in head-to-head battles in such places, ambitious entrepreneurs supported by local politicians could turn the tide against Bell.[15] Atlanta was one such case. In an agreement of 1906, North Georgia Electric, a power company, furnished electricity for the locally owned Southern Power and Light Company and the independent Atlanta Telephone and Telegraph Company. North Georgia was a cooperative venture started by local promoters to generate hydroelectric power at a nearby dam. The telephone company agreed to furnish space on its poles to bring this power to the city.[16] In the 1906 contract, all three private firms agreed to cooperate in the building of a transmission system, and to share the net profits of their operations. The city maintained final say over construction, placement of poles, and other technical matters. In this way, the municipal government and the local private interests planned and executed a comprehensive municipal utility system.[17]

This effort put Atlanta in a strong bargaining position with other utility service providers, particularly Bell. The city had a large market and, more importantly, a strategic one, due to its location along the main southern trunk line. The presence of a well-financed, city-supported competitor thus could not be dismissed lightly. Southern Bell agent A. G. Sharp recommended purchasing the independent telephone company and cooperating with the city.[18] Bell had its chance in 1907 when the utilities could not sell their bonds in the New York capital market and were desperately short of funds. Bell managers refused to take advantage of the opportunity, however, fearful of taking on the independent's heavy debt. Agent Sharp persisted in his recommendation, noting that despite the immediate conditions, the equipment of the telephone company was good and the electric utility properties were highly valuable. If local parties came up with the needed finance, they would profit handsomely and be hard to dislodge.[19]

As Sharp predicated, a citizens' committee organized and proposed to raise the needed capital when the panic abated.[20] At the same time Hoke Smith, whom Bell had opposed, was elected governor on an antimonopoly platform. While Smith at one time had been in the employ of Southern Bell, his political ambitions had turned him into a skillful and effective critic of the corporation.[21] The combination of a hostile state administration and united public and private municipal interests strengthened the Atlanta independent and damaged Bell's position. Receiving strong support from the city and outside manufacturers such as Stromberg-Carlson, the Atlanta Telephone and Telegraph Company continued in operation until 1918. Bell was able to purchase it only with assistance from state and federal regulators.[22]

While few New South cities possessed the enterprising spirit of Atlanta, other large municipalities were able to wield regulatory clout by using the law. All cities derived their authority to regulate telephone companies from their power to issue right-of-way franchises.[23] Holding over telephone firms the invaluable certificate to place lines and cables along or under municipal byways, they could control entry, rates, and service quality in the telephone industry, at least locally. No city in the South used this authority more aggressively than Richmond. In 1884 Richmond had granted to Southern Bell a franchise right-of-way in exchange for the firm's promise of a $10 per year limit on municipal government telephones.[24] After the resurgence of competition in 1894 it raised the stakes, demanding free service for government telephones. When SBT&T refused to comply, the city repealed the franchise.[25] Bell responded by taking the city to court, only to lose in a 1900 landmark federal decision.[26] The Richmond case established irrefutably the authority of municipalities to use right-of-way franchises to regulate telephone service.

The Richmond decision was especially frightening to Bell managers, for it revolved less around specific issues than the broad question of control. Disputes over rates and type of service could be negotiated, but Richmond's determined effort to control local service, backed by the court decisions, suggested to Bell that it could easily be ousted from important markets throughout the country by antagonistic city authorities. Richmond officials, keenly aware of their powers, had successfully challenged Bell over a basic issue—was telephone service a matter for public policy, or strictly a private concern. The court decided that telephones were a public matter, and left regulatory authority in the hands of city officials.

The Richmond case underscores the basic difference in attitude of Bell managers and city officials toward public utilities. In the eyes of city residents, the purpose of local government was to promote their welfare, putting local interests ahead of others. Most municipal governments followed this sort of policy in telephone regulation. They tried, for example, to protect their citizens' investments in independent telephone firms from Bell competition. For this reason, the towns of Spencer and Wendell, North Carolina, passed ordinances that granted Southern Bell a franchise only if it bought the property of defunct local telephone firms.[27] Rate policies that aimed to lower local costs regardless of the needs of the system were ground in a similar philosophy. The purpose was not simply to benefit individual consumers, but to promote local development through inexpensive public services. These ideas stood in direct opposition to the system-building outlook of Bell managers, who were trying to construct a national network that stretched beyond traditional political boundaries and local interests.

Still, strong as were southerners' commitment to local control, they often abandoned it when they were convinced that it was detrimental to long-run economic growth. The same logic that induced towns to regulate telephones to serve local needs prevented them from moving too far beyond Bell's orbit, however great their fear and resentment of "foreign" corporations.[28] Owners of a Jacksonville, Florida, independent firm, for example, failed in their appeal to gain city assistance in their fight against Bell.[29] Southern Bell President W. T. Gentry persuaded city officials that punitive measures against his company would frighten away other corporations and limit the city's options for telephone service in the future.[30] Fearing loss of needed capital and vital services, members of the growth-minded municipality came to terms with Southern Bell.[31]

Thus, while many conflicts abounded between Bell and southern municipal governments, opportunities for compromise also existed. After winning its case, for example, Richmond granted Southern Bell a new franchise.[32] Even big cities in the South conceded to Bell once they had obtained a few concessions. In smaller places, accord was even easier to achieve. Smaller towns took the most restricted view of telephony, as demand for service in such places was almost strictly local. Public officials, local telephone promoters, and important business elites in these places worked together closely to shape service to their needs. Yet Bell was often able to come to an agreement with small towns because their ability to resist the corporation was weaker than was that of larger cities. Rapidly growing communities in Florida and North Carolina valued telephone service too much to allow resentment against outside corporations to overwhelm their commitment to economic development. Through pragmatic negotiations with Bell, they enacted compromise contracts that suited local demands for service, but preserved Bell's goal of creating an integrated system.[33]

In 1903, for example, Southern Bell and Asheville, North Carolina, came to terms when the corporation bought the competing local exchange for $85,000 in preferred stock. Agreeing to keep rates to a yearly maximum of $40 for businesses and $24 for individuals for five years, Bell gained a favorable city franchise with no high taxes or requirements for construction such as expensive underground cables.[34] In Raleigh and Durham, Southern Bell formed a new type of organization which suited both its and the cities' needs. Combining forces with local entrepreneurs, SBT&T started the Capital City Telephone Company, which gained city approval to consolidate Bell properties and the toll lines of the formerly independent Interstate Telephone Company.[35] This arrangement allowed Southern Bell to extend its toll network into Raleigh and Durham while maintaining a semblance of local control.[36]

By 1912 Southern Bell had a clearly articulated policy of compromise with city officials. The corporation worked to bring municipal politicians, independent telephone promoters, and local businessmen, attorneys, newspaper editors, and civic leaders together in negotiations over telephone rates, service, and management. Through this broad-based political bargaining, the firm was often able to circumvent municipal regulations and limit city control.[37] Yet these compromises were often difficult to sustain. As Tampa, Florida, described its relations with Southern Bell, "it [is] the purpose of both parties to promote each other's interest whenever it is not incompatible with the interest of either party."[38] Rarely did the two interests coincide perfectly, leading city officials to wield their regulatory power. As Governor William Kitchen of North Carolina (a state of many, pugnaciously independent-minded towns) warned Southern Bell, his citizens would bow to no outside force. "If foreign corporations should unite in withdrawing from our state," he declared, "domestic corporations, obedient to the public will, respecting the people's law, will take [their] places."[39] Even when they compromised with Bell, southern municipalities took their autonomy and independence seriously, and stood ready to revoke their franchise grants if necessary.

Bell encountered this same attitude throughout the nation, for as its system grew it ran up against durable pockets of protected local interest. Thus conflicts in Bell-municipal relations were not unique to the South, but part of a nationwide response to the emergence of large-scale systems. Indeed, in many other regions of the country, splits between Bell and city governments were even worse than in the South. In the Midwest, fiercer competition in the telephone industry gave city regulators more power than their southern counterparts. Whereas in even large southern cities such as Richmond, city officials followed the interests of a narrow local elite, midwestern municipalities took a broad view of local needs, and operated with reference to a strong tradition of public utility regulation. In contrast to the South, where municipal regulation aimed at bread and butter issues like low cost and free government service, midwestern cities pursued a complex regulatory policy. Such efforts presented even greater challenges to the Bell system.

Even before the advent of competition, Chicago had used its franchise power to demand payment of a 3 percent tax on gross telephone receipts and free municipal government service.[40] Since Chicago was the major juncture of the important midwestern long-distance system, Chicago Bell had little choice but to agree. The city's leverage increased after 1894, as Bell and independents vied for control of the crucial market. In 1907 the city granted Bell a new right-of-way franchise, but got in return expanded free municipal service.[41] Not content with these concessions, Chicago in 1913 acted on a clause in the 1907 contract that permitted it to set a ceiling on city rates and limited charges to five cents per call.[42]

Although Chicago authorities wanted low telephone rates, they used their power in innovative ways to achieve other goals as well. The city had reserved for itself in 1907 the right to set up a municipally owned telephone company.[43] Though never exercised, this option gave Chicago another lever to use against Bell. Taking advantage of its leverage, the city in 1913 started a municipal telephone bureau to gain access to important financial and business information about Bell operations. Officials also set up new classes of service. Taking upon themselves the planning of municipal telephone service, they arranged for special low-cost neighborhood exchanges with limited outside connections.[44]

Throughout the Midwest, other municipal authorities used their regulatory authority in similar ways to shape telephone service to suit their needs. Like Chicago, Detroit and Indianapolis passed ordinances regulating rates and reserving for themselves the option of taking public control of city telephone service. These policies left Bell fearful that municipal governments would begin "picking off the largest exchanges one by one, and leaving AT&T with the less important territory."[45] While this worst-case scenario did not materialize, municipal control in the Midwest, combined with competition from independent service providers and equipment manufacturers, did seriously threaten Bell's ability to operate its system autonomously.

Bell could not combat the strong municipal regulatory movements in the South and Midwest through private means alone. With the legality of the right-of-way franchise unassailable, the company had no means of easily circumventing city control. Calls for a federal law granting telephone firms access to rights-of-way similar to one passed in 1866 for telegraph firms had not been successful; and some members of the Bell organization feared that such a law would do more harm than good by providing competitors with easy access to municipal streets and byways.[46] The company continued seeking franchises city by city, compromising with municipal authorities where possible; but over time this approach became increasingly ineffective, as cities in the South and Midwest exercised their powers more aggressively to serve local interests.[47] Nor did private acquisitions of competitors offer a way out. Aggressive acquisition raised the specter of an antitrust suit, something that E. J. Hall's compromise policies had sought to avoid. Unable to bypass municipal regulators on its own, Bell searched for public sanction for its policies from other quarters.

Between 1910 and 1920, Bell found this support in state legislatures and public utility commissions.[48] In these years, state commissions took over regulatory responsibility for telephones from local governments, as they did in many public utility industries.[49] Though not captives of the corporation, state governments tended to agree more with the ideas of telephone service put forth by Bell managers than those espoused by city officials.[50] They viewed municipal dealings with telephone firms as either irresponsibly procompetitive, ignorant of the principles of rational economic management, or simply corrupt. Often state-level regulators charged city governments with all three sins, condemning them for fostering economic waste and inefficiency.[51]

None of these charges against the cities was wholly accurate. The thrust of municipal policy was not to increase competition. In 1902, for example, Southern Bell reported that no opposition exchanges were under construction, and that existing ones were being acquired in private deals, often with approval from city officials.[52] Though clearly overly optimistic, the report nonetheless spoke some truth. Direct competition was ceasing to be a major issue. While at times municipal governments used their powers to foster competition by granting multiple franchises, most of them quickly realized that local service was naturally monopolistic. They instead used the threat of competition to protect themselves against monopoly exploitation and to achieve other ends, favoring with a franchise whichever telephone firm agreed to serve local interests best.

State authorities misconstrued municipal regulatory intentions in a number of other ways as well. They charged cities with corruption because they saw that many of them were extracting free service and rebates from telephone carriers and demanding other concessions in exchange for a franchise. But these actions conformed to the cities' belief that telephone service should promote community interests and boost local development, a conception of public policy that pervaded municipal regulatory practice, as we have seen.[53] Only in the twentieth century, with the rise of independent commissions and the growth of professional social science, did regulation lose its ties to the community. When it did, local interests were sacrificed to those of the state or larger corporate bodies. It is not self-evident, however, that the common good was served by this shift in public policy.

Accurate or not in their assessment of municipal regulation, state legislatures began to deprive local governments of power over telephones. By 1910, many had granted final say over the industry to independent commissions. Taking a broader view of telephony than the municipal governments, these bodies broke down local constraints and gave freer rein to the Bell system. In doing so, they helped to solve the crucial problem of technical compatibility among regional telephone systems. Through their rulings, Bell's standards of service gained backing in state after state. With this support, the corporation was able to override local concerns and join all regions of the country into one system.

In the South, state telephone regulation developed more slowly than in other places, despite the numerous conflicts between Bell and cities. Though Louisiana, Mississippi, North Carolina, South Carolina and Virginia were the first five states to formally place telephony under the supervision of commissions—generally railroad commissions—these bodies initially intervened very little in the industry. They continued instead to devote the preponderance of their time to railroads. Nor did they possess much real power over telephone service or rates, despite the letter of the law.[54] Before 1920, for example, southern commissions could not issue certificates of public convenience and necessity (though many in the North and West could).[55] Like municipal franchises, certificates governed entry into the industry. With them regulators could decide which telephone company could serve which market and how many firms could operate in a given area. If necessary, certificates could also be used to force carriers to interconnect to create a single telephone network. Southern state commissions' authority, however, stemmed from general laws giving them control of all common carriers. It was not clear if telephone companies were common carriers. Common carrier law, moreover, did not address the key issues of the telephone industry—service quality and system design. Rather, it spoke to an older set of concerns, such as fair competition, nondiscrimination, and company financial stability.[56]

Lacking substantive power, southern state commissions initially exercised little control over city governments. Despite its mandate to oversee telephones, the Georgia Railroad Commission early on deferred to municipal authorities and refused to contravene their decisions.[57] In North Carolina, state regulators had only a supervisory role, reviewing and approving decisions made at the city level.[58]

Yet despite their handicaps, several southern state commissions did find ways to assume control over telephony. Though before 1911 Florida laws governing telephones were general and weak, the Florida Railroad Commission made the most of what it had.[59] It decreed that its rulings superseded the decisions of municipal governments, and asked for and got more specific wording in the sections of the laws covering telephones, thus resolving in its favor the question of ultimate authority.[60] With its power over municipalities confirmed, Florida's commission was able to advance a regulatory philosophy much in keeping with the goals of Bell managers. It enforced laws that demanded equal charges for like services and allowed no preference for particular localities or customers.[61] Though these laws were originally intended to end discriminatory practices by carriers, the Florida Commission applied them to the telephone industry so as to prohibit the granting of free service to municipal governments, thereby reducing their bargaining power with Bell. By 1913 the Florida Commission was able to void all rates "insufficient to yield a reasonable compensation for services rendered."[62] Enacted to curb "cutthroat" competition, such stipulations also favored Bell's systemwide approach to prices and service. Bell managers for years had argued that competition, free service, and municipal efforts to tie prices to local costs hurt system growth; now their position was finding a sympathetic ear among state regulators.

The Florida Commission also acted decisively to shape telephone technology in accord with Bell needs. In this area, even more than in rate policy, it fostered a broad conception of telephony, giving the national system preference over local demands. State regulations prohibited the use of facilities which were "inadequate, inefficient, improper, and insufficient."[63] Though such rules could have been used to protect consumers, in the manner the Florida regulators employed them they hurt independent firms. Small-scale carriers had achieved a measure of success against Bell by creating a simpler, cheaper type of telephone service attractive to small towns. The commission's rulings, however, tended to undermine this price advantage. The state regulatory body ordered a Daytona independent to upgrade its service to provide direct line connections to all who wanted them, forcing the company to install expensive common battery equipment. It fined the independent Mariana Telephone Company $100 for discontinuing an unprofitable toll connection, hurting the firm in its competitive battle with Bell.[64] And it demanded that all telephone firms provide continuous service, a ruling that spelled certain death for many small and marginal carriers.[65] By 1917, the Florida Commission's definition of adequate telephone service matched Bell's. The regulatory body banned the use of non-Bell equipment on connecting lines, a policy much in accord with the national firm's attempt to standardize equipment among its sublicensees.[66]

While many of the Florida Railroad Commission's rulings favored Bell, it was not captured by the telephone giant. The state regulators also set equitable toll schedules, and demanded interconnection between Bell and non-Bell firms when in the public interest.[67] Yet in the way that even seemingly neutral laws were applied, one can see the predisposition of the state to favor Bell-type service. In the hands of progressive-minded state regulators, traditional regulatory concepts such as antidiscrimination and just and fair rates, concepts not intended to favor one company over another, ended up reducing municipal control of telephones and sanctioning Bell rates, service quality, and technology. Each of the Florida Commission's rulings was premised on a broad national, rather than narrow local conception of telephone service. In this way, the body served Bell's needs.[68]

This pattern in telephone regulation was most pronounced in economically developed southern states, where the Bell system was relatively more advanced. While the Florida Railroad Commission had to stretch its powers to achieve its ends, commissions in more industrial southern states tended to receive strong powers over telephone firms directly from the legislature early on. In populous

Virginia, a state-level commission quickly gained substantial authority over telephones. A 1904 law gave the newly created Corporation Commission control over all public service corporations and spelled out specific rules for telephone service.[69] The body was authorized to limit competition and prohibit discriminatory practices.[70] With these powers, it was able to reduce municipal influence over telephony by denying cities the power to extract rate concessions from carriers. In 1914, the Virginia board also received formal rate-making authority and the right to review municipal ordinances and contracts with telephone firms.[71]

Georgia, the hub of the southern toll network, also gave its Railroad Commission clear powers over telephones.[72] In a 1907 statute the legislature put to rest nagging legal questions that had constrained other southern regulatory commissions. Courts tended to strike down commission telephone decisions when the law did not specify powers over telephones directly, but instead simply mentioned all common carriers.[73] The new Georgia law freed the Railroad Commission from this worry. In 1910, the commission moved to reduce municipal power over telephony by ruling in the Dawson Telephone case that its regulations superseded municipal ordinances and decrees. This important decision was upheld by the state Supreme Court, making the Georgia Railroad Commission the primary regulator of telephone service in the state.[74]

With its power over municipalities secured, the Georgia regulatory body took up the important issues of telephone rates, service, and technology. It ruled that provision of service at prices below cost was a discriminatory practice. This decision prohibited concessions such as free or reduced municipal service, thereby limiting municipal leverage over telephone firms and reducing price competition.[75] Bell benefited from such actions, for its high-quality system frequently required rates in excess of what municipalities demanded.[76] Addressing matters of service quality and technology, the Georgia Commission also followed policies in accord with Bell needs. Once again, not all of its rulings directly or obviously favored the corporation. Sometimes the Georgia body simply approved compacts worked out by Bell and individual cities; other times it placated consumers by tying rates directly to improvements in service.[77] But when necessary, the commission also foisted on localities a type of service they did not seek, ignoring the pleas of rural consumers and the ordinances of municipal authorities.[78] In one case, the commission put an end to inexpensive party-line service, declaring it "inefficient."[79] Such a decision was clearly a victory for Bell, which sought to eliminate these types of low-profit, low-quality service whenever it could without incurring the wrath of consumers. The commission ruling enabled it to do so without fear of reprisal.

Bell was fortunate that in states such as Georgia and Virginia independent commissions quickly and decisively assumed regulatory authority over municipalities. Georgia was a key state in Bell's southern toll network; municipal regulation, with its strong orientation toward local concerns, could have significantly slowed the Bell system's progress. Virginia, the wealthiest state in the seaboard South, also had a strong demand for long-distance telecommunications. There too the potential costs to Bell of municipal dominance were great. Not surprisingly, therefore, Southern Bell executives approved of the new regulatory laws of Virginia and Georgia.[80] As in the case of Florida, however, Bell's involvement in state politics and its efforts to shape regulatory practice seem to have been small, limited to minor rhetorical efforts. Shared values were responsible for bringing public officials in these states in line with Bell interests.

Taken with the progressive-era faith in the superiority of large-scale systems, politicians in economically advanced southern states worked to expand the Bell system and root out the vestiges of localism that seemed to slow its progress.[81] As Georgia railroad commissioner C. Murphey Candler stated, his job was to protect both corporations from an ignorant, greedy, and narrow-minded public, and the public from avaricious corporations.[82] This could best be accomplished, Candler believed, by an independent commission. Corporate growth tended to eliminate traditional community standards of social value and status. New organizations such as the Railroad Commission replaced those lost standards by regulating relations between the public and modern corporations in a way that was most efficient for society as a whole. From this perspective, Candler and his colleagues were much closer to Bell than locally minded-municipal governments.

As time went on this regulatory philosophy gained strength

throughout the South. Between 1912 and 1914 most southern states granted commissions greater powers over telephones, allowing them to supplant municipalities as the primary regulators of telephone service. In state after state commissions decreed that competition could "[never] be a consistent and efficient regulator of rates and other conditions in a local utility field." [83] Few city governments had been explicitly procompetitive, as we have seen. But the rhetoric of efficiency served its purpose: It reduced local control of telephones and propagated a new, progressive style of regulation. These regulatory efforts conformed to Bell's own plans for building up toll service in profitable areas and bringing more territory into its network.[84]

Still, even by the late 1910s this regulatory philosophy was far from universal in the South; and it was much weaker than in northern states. The South's traditions of local rule were simply too strong—and correspondingly, demand for Bell system service too weak—to allow progressive-minded state legislatures to override all municipal authority. North Carolina, for example, continued to have a very weak state commission into the twentieth century.[85] Formally, telephones were placed under the state's Corporation Commission when it superseded the old Railroad Commission in 1897. The Corporation Commission was charged with responsibility for monitoring telephone and telegraph companies, and was authorized to set "just and reasonable" rates.[86] But the commission proved ineffective against city governments, despite its formal powers.[87]

The North Carolina legislature had significantly weakened the Corporation Commission's ability to control telephones by sharply distinguishing the parts of the new law covering railroad and transportation companies from the parts covering other carriers. In the railroad sections, it included specific rules regarding rates, discrimination, and anticompetitive practices.[88] Elsewhere, the lawmakers were far more vague, providing only general regulatory guidelines. For industries such as telephony, the act did not address key issues such as rate discrimination, rebates and drawbacks, or interconnection.[89] North Carolina courts refused to apply the vague sections relating to non-rail carriers, reducing considerably the commission's reach in the telephone industry.[90]

This tradition of weak state-level telephone regulation followed North Carolina into the next decade, despite amendments in the original corporation commission law. In 1907 the legislature empowered the regulatory body to make rules and orders for the "comfort and convenience" of public service corporation patrons.[91] In the same year, it extended commission supervision to all firms or individuals owning, operating, or renting telephones. Finally, the lawmakers applied to telephony all rules and regulations pertaining to railroads.[92] These changes, however, were far less significant than the amendments in telephone law carried out in Georgia and Virginia at the same time. They seem to have been passed mainly with the intent of allowing the commission to investigate consumer complaints, rather than permitting the body to govern rates, competition, or service. Missing in North Carolina was the specific sanction present in other southern states that would have enabled the Corporation Commission to bring its power to bear against municipalities. Recognizing the narrow scope of its authority, the commission dismissed cases as out of its jurisdiction, rather than test the limits of its administrative reach.[93] The furthest the body would go was in approving private deals worked out by Bell and cities.[94] Under these conditions, issues vital to the fate of the Bell system remained in local hands.[95]

North Carolina's Corporation Commission had more to say on telephone rates, though on this issue as well its action was indecisive. Defeated by the courts in an 1899 effort to set maximum prices for telephone service, it tried a different tack in 1906, adopting a rough value of service pricing policy.[96] The commissioners decreed that a "uniform rate for telephone exchange service in North Carolina, regardless of the value to patrons or the amount invested, would not be reasonable or just."[97] On this basis, they later let Southern Bell increase its rates in Wilmington, noting that its 4.3 percent return on investment was too low. But the case did not go wholly in Bell's favor. The commissioners also ruled that their decision was based strictly on *local* conditions, stating "the petitioner [SBT&T] is entitled to a reasonable profit on the amount invested in the exchange service at Wilmington."[98] If they adhered to this logic, the commissioners would ignore costs common to the whole system and costs that could not be easily assigned to any one place. Failing to take into account these joint or system costs, they would set the rate base too low to guarantee the Bell system an adequate return. Construed in this way, their rulings on prices seemed to reinforce local over systemwide concerns.[99]

As these variations among southern states suggest, the South was torn between its desire for modern economic institutions and its strong aversion to alien control and ownership of vital services. In the more industrially advanced states of the region such as Georgia and Virginia, Bell received solid support from independent commissions to pursue system building. In other places such as North Carolina, where Bell control of the market was weaker, demand for its long-distance service less robust, and competition from independents stronger, state commissions moved far less decisively to support the Bell system over alternatives. Falling between these extremes were "frontier" states such as Florida. Though developing rapidly in the north, Florida still contained many isolated towns and rural areas. This uneven pattern of growth perpetuated "a hostility to corporate enterprise," which creative action by the state public utility commission could only partly overcome.[100]

Ambivalence of this sort tended to restrict the effectiveness of public policy in the South. While weak commission regulation in North Carolina helped to preserve telephone competition in the state, it also narrowed the scope of public action to strictly local issues and failed to address the possible advantages of the Bell system to consumers. In a world coming to be dominated by large-scale, national organizations, a retreat to localism was a move of weakness. On the other hand, in those states with strong state commissions, policy was no more creative. Reacting against deep traditions of local control, progressive-minded public officials in Georgia believed that they were embracing a policy that could preserve the fruits of the Bell system while guarding the public good. As we shall see, however, in comparison with other places their efforts were one-sided and simplistic.

The South was not, of course, the only region torn in this way. Wherever the pattern of small towns, strong agrarian traditions, and independent telephone firms existed, people resisted incorporation into the Bell system. In each case state regulators played a key role in moving these places into that system, overcoming the resistance of small-town entrepreneurs and municipal governments. In the Midwest as well as the South, state congressmen viewed local control as a backward and often corrupt effort to extort concessions from foreign corporations. There too they moved against it by empowering independent commissions to oversee telephony. Midwestern Bell managers eagerly awaited this change in the law, believing that the "only hope" against aggressive municipal governments were state-level regulatory commissions.[101] As in the South, these commissions saw themselves as nonpolitical decision making bodies free from local concerns and capable of making rational, objective policy.[102] Generally their policies were defined in ways which conformed to Bell system needs. Nonetheless, differences in the telephone industry and political traditions of the South and Midwest led to different outcomes in public policy.

In contrast to the South, state telephone regulation in the Midwest evolved in a much more volatile political climate. Aided by fierce competition and a strong tradition of home rule, midwestern municipalities held fast to their powers over telephone companies.[103] Midwestern cities also tended to be more ambitious in their policies and more politically skillful in their bargaining with telephone firms than their southern counterparts. Chicago, as we have seen, exerted great influence over telephony because of its strategic location. At first the state legislature could not loosen the city's grip on telephones. Bills were introduced into the Illinois Senate aimed not at replacing municipal with state-level control, but encouraging competition by opening up the Bell toll system to independent firms.[104] Though none of these early proposals became law, they demonstrated the strong sentiment against the Bell system even at the state level.[105]

Over time, these sentiments waned, and the move to commission regulation of telephones grew stronger. In 1913, the Illinois legislature created a Public Service Commission (PSC) to oversee many public service corporations. Yet even then the spirit of local control remained. Some members of the Illinois Senate tried to placate powerful Chicago interests by dividing the PSC's power between that city and the rest of the state.[106] This unsuccessful proposal would have left downstate towns under commission control, while allowing Chicago to regulate its own utilities. Other senators, also amenable to local interests, proposed bills granting all towns the right to set their own rates.[107] Though the state PSC retained most of its power, the issue of local control returned again and again. In 1917, a move began to restore home rule of utilities, and in 1920 Leo Small successfully fought his gubernatorial campaign on a promise to abolish the commission.

Although continually in the midst of controversy, the Illinois Public Service Commission still had a strong enough legal mandate to exert some authority over telephones. It was able to reduce the power of municipal governments by banning rates below tariff.[108] Like commissions elsewhere, the Illinois body supported large-scale systems over local options. Early on it declared that it would allow no competitive entry or duplicate service in telephony. Direct competition was not the main issue, of course, as Bell reports showed that no more than 25 percent of the state had duplicate service.[109] As the commission quickly made clear, however, its anticompetitive policy was designed to eliminate divisions and differences in the telephone service of different places. By limiting entry, sanctioning Bell acquisition of independent firms, and permitting the consolidation of smaller companies into larger ones, the commission reduced local options and forced all places to conform to a single standard of service.

The Illinois regulators also worked to standardize service by bringing more competing telephone carriers into Bell's growing national network. Though it could not force mergers, the body threatened in 1915 to order interconnection between separate firms to unify telephone service.[110] Bell had already agreed, in its 1913 compromise with the United States Justice Department, the Kingsbury Commitment, to allow certain independent firms access to its system.[111] Midwestern independents, however, resisted connecting with Bell, fearing that they would be "swallowed up" by the telephone giant. They noted, correctly, that in Bell's hands interconnection through the sublicense contract was actually a competitive weapon.[112] Yet because interconnection offered compensating advantages to the smaller competitors, the Illinois Commission was often able to tip the balance and get them to agree to join the Bell network. In one case, for example, it used interconnection as part of a compromise to help consolidate telephone service. Earlier, Bell had tried to buy the independent Interstate Telephone Company, raising a loud cry of protest from small non-Bell firms. Interstate provided these carriers with crucial long-distance connections, and they feared that if the company fell under Bell control, they would lose this service. The commission authorized the takeover, however, though it guaranteed non-Bell firms access to outside lines.[113] In this case, interconnection proved to be the independents' only option.

Used in this fashion, interconnection conformed closely to the policies AT&T had devised to combat competition. The independents gained access to the Bell system, but at the cost of their autonomy. Though they were not driven from the industry altogether, they were not allowed to form competing long-distance companies either. When they stuck to their own markets and did not challenge the national firm, they were rewarded with access to its network. This policy helped to expand the scope of the Bell system, but it also narrowed the range of service options available to customers.

As interconnection and consolidation drew more areas into the Bell system, enforcement of standards for service quality became a prime regulatory concern. This issue proved more difficult to settle in the Midwest than in the South. In the latter region, state commissions simply adopted Bell standards for service and equipment.[114] But in the Midwest, greater disparities between markets and a larger variety of users demanded more complex policies. The region's many independent firms served a wide range of customers, particularly in the rural and outlying regions beyond Bell's reach. To make these customers part of its system, Bell either had to upgrade the quality of their service or lower its own standards. Increasing service quality often brought angry protests over higher rates. For many years, therefore, the firm had simply made little effort to reach such customers. State regulators took a different approach, formulating policies that placated these customers while maintaining Bell system standards.

To keep all areas united in one telephone system, Illinois's state commission prevented telephone companies from withdrawing from less profitable areas, ordered the construction of rural lines, and mandated the incorporation of outlying areas into larger systems.[115] By requiring firms to serve marginal areas, however, the commission left itself no alternative but to devise a new rate policy. If large carriers like Bell were forced to make new connections at rates customers could afford, they would lose money. The Illinois Commission responded with statewide rate averaging.[116] With averaged rates, high profit areas subsidized low profit ones, and all areas conformed to a single standard of service. A prescient move on this commission's part, rate averaging would soon become a national policy, helping to solve throughout the Bell system the nagging problem of bringing service to remote areas.

Through interlocking rules regarding rates, interconnection, and service quality, the Illinois body devised an innovative regulatory alternative to the policies of individual municipal governments. City authorities had used competition for franchises to match telephone service to local needs. As a result, service quality differed in different markets, undermining Bell efforts to create a unified national system. State regulation fostered compatibility between markets, eliminated price competition through rate averaging and "antidiscrimination" rulings, encouraged interconnection between carriers, and standardized telephone equipment.[117] The result was a single type of telephone service in the state, though one sometimes at odds with the needs and desires of individual towns and cities.

In both the Midwest and South, state-level regulation of telephones evolved out of the severe conflicts between local governments and the Bell system. Not surprisingly, where those conflicts were not severe, a somewhat different regulatory pattern emerged. Such was the case in New England.

In this region, even remote towns largely accepted Bell telephone service. Early on, municipalities in Maine, New Hampshire, and Vermont, though they shared many of the economic and geographic features of the Midwest and South, had acquiesced to the values of the Bell system.[118] Their position changed little with the rise of competition. Some places, resenting Bell's monopoly power, tried to use competition to prevent exploitation by the corporation; others occasionally demanded concessions such as free service in exchange for a franchise.[119] But most remained quiescent. In Manchester, New Hampshire, businessmen ignored the pleas of some citizens for greater competition in telephony, declaring dual service inefficient.[120] Nashua, New Hampshire, rejected a petition by an independent promoter to start a telephone company in the town to link up with exchanges in nearby Lowell and Haverhill. The city stuck with Bell service.[121]

Businessmen and entrepreneurs in New England as well saw little need to challenge Bell's dominance of telephone service. In 1895 Edgar Gray, an independent telephone promoter, arranged to have his toll lines, which served several small towns in New Hampshire and Vermont, linked to Bell lines. This voluntary compromise pleased New England Bell, the municipal governments, and Gray himself.[122] In 1899, another promoter, F. B. Cross, sold his People's Telephone Company to Bell. Satisfied to leave the business with a modest profit, he believed that consumers would be better served by the national firm. In this same spirit, A. C. Brown readily sought compromise with Bell, declaring that in the long run a single unified network was best for all consumers.[123] By 1911 a strong sense of cooperation had emerged in New England's telephone business. At a convention of Bell sublicensees, the directors of connecting companies agreed that they, New England Bell, and AT&T all had responsibility for different parts of a single system.[124] Cooperation among members made this joint venture possible.

This broad consensus on telephone service in New England stemmed directly from economic conditions in the region. In even New England's most rural and remote sectors, the level of development was high, exceeding that of most of the South. Bell's toll network was also more extensive in the Northeast than elsewhere, surrounding and isolating independent firms and giving them little opportunity for forging their own systems.[125] Under these conditions, New England entrepreneurs, municipal governments, and consumers quickly developed a strong appreciation for the large-scale system Bell was introducing. Local concerns simply did not dominate the minds of New Englanders as they did those of people elsewhere. Thus, though New England municipalities had the power to regulate telephones, they rarely used it in ways injurious to Bell's interests.

Since local constraints did not interfere with system growth, state commissions in New England only gradually assumed power over municipal governments. Early on, states in the region did give utility commissions formal jurisdiction over telephones.[126] But these laws were weak and did not reflect a strong interest in state-level control. Maine's legislature, for example, refused to grant the Public Utility Commission power to overturn municipal ordinances.[127] Vermont's laws were even weaker. Though the state Public Service Commission was given statutory authority over telephones and telegraphs in 1908, its powers were general. The law made clear, moreover, that towns retained control of the physical placement of poles and wire.[128] Before 1915, Vermont's PSC lacked the authority to review corporate charters to see if they were in the public interest.[129] In 1917 it did receive full power to issue certificates of public convenience and necessity, allowing it to regulate entry into the telephone industry.[130] Nonetheless, the state's regulatory laws did not specifically address conflicts between commission rulings and municipal decrees, leaving lines of authority between state and local governments vague. Elsewhere in New England, other issues involving municipal power, such as the right of cities to extract free or reduced priced service from carriers, were hardly addressed.[131]

Northern states, unlike those of the South and Midwest, also at first did not try to restrict competition in the telephone industry. Maine's Public Utilities Commission of necessity adopted an explicitly procompetitive policy. Before 1913, the state legislature prohibited it from using its powers to control entry.[132] Passage in 1913 of a new public utilities commission law reflected a continued ambivalence of policy. Though it favored monopoly as a means of increasing service efficiency, parts of the law also aimed to protect consumers and local capital from powerful corporations.[133] Vermont's 1908 public service commission law also prohibited restrictions on entry or limitations on competition.[134] Despite feelings by some board members that this clause should be repealed, it remained in effect until 1933. New Hampshire was only slightly more decisive, taking a stand against competition in 1914.[135] In many cases, however, its utility commission was unwilling to determine who could provide telephone service, or under what terms it should be offered.

In their unwillingness to restrict entry into telephony and other public services, New England legislatures were responding in part to their states' strong antimonopoly traditions. Though some northeastern lawmakers extolled the virtues of large-scale systems, others restrained their exuberance with reminders of the dangers of monopoly. In Vermont, for example, the governor hesitated signing the 1909 public commission law, fearing that the new body would only foster monopoly.[136] To allay such fears and meet demands for consumer protection, Vermont's PSC reviewed the rates and service quality of almost every telephone firm in the state. Where it found small companies operating in restricted territories, it authorized special low rates.[137] Fearing Bell's market power, it also demanded reduced charges for Western Electric equipment. Finally, unwilling to permit Bell to operate with autonomy, it investigated the firm's Vermont operations and ordered it to reduce its depreciation reserves from 5½ percent to 4 percent in an effort to tie rates more closely to the value of the company's property.

New England's combination of a strong antimonopoly persuasion and a broad consensus on the value of Bell service did not always fit together neatly. While state commissions in the region, like those elsewhere, wanted to expand the scope of the Bell system, they also had to pay attention to citizens' fears that the corporation's growing size and power would lead to abuses. For this reason, New England's state commissions, more than those elsewhere, addressed concerns such as overcapitalization, stock watering, and corporate corruption, issues that faded into the background in the South and Midwest, where conflict between Bell and municipalities was paramount.[138] Conflicts in New England thus arose not as Bell and local interests fought over the shape and design of telephone service, but as public officials sought to protect consumers from the untrammeled power of the giant corporation. To meet these goals, New England state commissions had to devise sophisticated regulatory policies.

Often New England's basic consensus on the value of the Bell system provided the means of resolving these conflicts of policy. States in the region began to move away from an explicitly procompetitive stance, reinforcing Bell's control of the northern market.[139] As this transpired, commissions quelled public distrust by reminding consumers of the efficiency of a single, large-scale system.[140] At the same time, however, they were careful about regulating that system in the public interest. Mandated to protect consumers, Maine's commission took seriously the complaints of small towns that felt they were being discriminated against by Bell.[141] The regulatory body allowed, for example, the Franklin Farmers' Co-Op Telephone Company to furnish service in New Vinyard, even though another carrier already served that market. While appreciating the problem of dual service, the PUC nonetheless maintained that the co-op brought together many places not served by Bell, and linked rural areas with the town.[142]

Through such policies, the commissions of New England, like those in the Midwest, helped to expand the Bell system in ways which took account of the special needs of residents on the margin. The Maine PUC, for example, refused to permit NET&T to acquire the independent Monroe and Brooks Telephone Company in 1919, noting that the company provided a distinct type of low-priced service of great value to patrons in its service area.[143] For similar reasons, the Vermont PSC forced New England Bell to reinstate four-party line service favored by St. Johnsbury residents, refusing to accede to Bell's argument that the service was "inefficient."[144]

At times, the New England commissions, in their effort to expand service, fostered policies that favored the Bell system over its competitors.[145] Maine's PUC sided with urban-based telephone companies over rural independents, forcing farmers to join their lines to more expensive, high-quality municipal systems.[146] It favored consolidation as well, insisting that large companies acquire small, marginally profitable firms to maintain adequate service in outlying areas.[147] In a similar manner, New Hampshire regulators encouraged consolidation as a means of providing more efficient service and eliminating wasteful competition.[148] As time went on, the ideology of the system, generally the Bell system, began to dominate New England state commissions, and the regulatory bodies strove to eliminate local variations in service quality and rates.[149] Nonetheless, like the commissions of the Midwest, those of New England blended their strong belief in the value of a single system with other regulatory concerns to form a complex telephone policy that served a variety of interests.

By World War I, telephone regulation throughout the

United States was stabilizing around a common set of principles. In all regions, state utility commissions, working out of shared values with Bell, promoted the large-scale telephone network over local alternatives. Bell reinforced this tendency by abandoning its antagonistic stance toward state and federal authority, and by agreeing to accept regulation of its near monopoly. While AT&T president Frederick P. Fish in the early 1900s had scoffed at the idea that the company needed assistance from regulatory authorities to build its system, this attitude quickly changed.[150] In accord with the new philosophy of compromise developed by E. J. Hall and Theodore Vail, AT&T endorsed state regulation.[151]

This same attitude filtered down to the lower echelons of the Bell organization. Initially Southern Bell managers had feared the rise of state utility commissions, despite the firm's grave conflicts with municipalities.[152] By 1909, however, Southern Bell president W. T. Gentry reversed his position and was declaring state regulation a boon to his firm and the Bell system.[153] Eventually even Bell's rivals agreed. In 1919, O. F. Berry, general counsel for the Independent Interstate Telephone Association, a body which had been dedicated to fighting Bell dominance, endorsed state-level regulation.[154]

Bell clearly benefitted from state regulation. As Bell managers came to realize, without the work of the commissions, municipal governments would slow the growth and restrict the reach of their system.[155] State regulation helped to solve other important problems connected with Bell system growth as well. It brought together disparate sections of the market and solved long-standing controversies over rates, service quality, and standards, generally in Bell's favor. In this sense, state regulation was the final part of the policy initiated by E. J. Hall to deal with regional variations in economic conditions across the system. By denying small firms municipal assistance, state regulators reduced their market and thereby closed off local options. By sanctioning Bell acquisitions and helping to consolidate the industry, they forced smaller towns and rural areas to conform to Bell system standards. And, by fostering rate policies which eliminate low-priced local service, they cut out this alternative to Bell service.[156] These policies reduced the local opposition that had been present since the earliest days of the Bell system.

Federal power also played a role in this process, though a somewhat smaller one. Through the threat of an antitrust suit, the Justice Department was able to force Bell to make certain concessions on its interconnection and acquisition policies. In the Kingsbury Commitment, Bell agreed to allow independents access to its system and to refrain from acquiring competitors without government approval. But Kingsbury was not to be used to foster renewed competition. In this sense, it fit the general pattern of regulation emerging in the states. Bell allowed interconnection, thereby increasing the scope of its system in remote and marginal areas served by non-Bell firms; but it did so only when these firms conformed to Bell standards. Similarly, the corporation ceased buying independents with which it was in direct competition, though it was still free to acquire or sublicense others.

Political power, then, played an important role in the growth of the Bell system, as it has in the creation of other large-scale, technologically complex systems. In the electrical power industry, for example, regional variations also led to conflicts of interest between groups, and fostered different values toward technology and business enterprise in different places. As Thomas Hughes has shown, in some cases local political and business interests joined forces to create a particular technological style that blocked large-scale system growth. In other cases, regional economic and political elites by-passed private corporations and constructed their own large-scale enterprises.[157] In telephony, state regulation, oriented towards large-scale enterprise, overrode local interests to advance a system in line with what Bell, the largest firm in the industry, wanted.[158] Bell and the government regulators, separately but in concert, prompted a new concept of telephone service. Through their efforts, an idea that had been hotly disputed a decade earlier—the rationality, efficiency, and inevitability of an interconnected national network—became a commonplace. In this way, they forged America's privately owned, publicly regulated national telephone system.

The effects of this regulatory change on the economy of the

South were mixed. By assisting the completion of the national telephone network, southern commissions helped to strengthen the links between their states and the national economy. In this way, they contributed to long-run growth and development. Yet the resulting public policy in the South was not very innovative. It failed to address the legitimate needs of many consumers at the margins of the industrial world. In the Midwest and Northeast, by contrast, regulatory policy served a variety of interests while aiding the growth of the Bell system. Midwestern commissions used innovative policies such as rate averaging to bring all parties into the Bell system. Clearly a compromise, this pricing arrangement placated consumers in outlying and rural areas, while permitting Bell to earn an adequate return on its investment. New England too developed a complex set of public policies, though the region started from a different point than the Midwest. In the North, antimonopoly sentiments and a pro-Bell consensus, existing side by side, led to greater protections for consumers against the emerging Bell monopoly, even while other policies protected that monopoly. In both the Northeast and Midwest, sharp bargaining with the Bell system produced a growing telephone network that met at least some of the needs of a wide range of customers.

The South appears to have missed this opportunity to bring public power to bear in creative ways against the Bell corporation. In part, the problem stemmed from economic conditions in the region. Independent telephone firms put up a surprisingly weak fight against Bell, thus lending little aid to municipal governments. Municipal governments themselves, even in large cities such as Richmond and Atlanta, were often willing to compromise with Bell for a fairly modest price. For all their anti-Bell bluster, municipal authorities paid scant attention to the needs of the larger society. This seems to have reflected the limited distribution of telephones in the region. Telephony was mainly an elite concern, and southern town elites worked mainly in their own interest. They guarded their investments against Bell incursions, but were willing to concede a free hand to northern capital for the promise of development. They showed little interest in the needs of the majority of poor and rural consumers. Thus, the persistence of local control in states like

North Carolina, rather than stimulating cohesive community action and interest group bargaining, only increased community isolation and insularity. While protecting local interests, this strategy ultimately left communities weaker in dealing with powerful national corporations like Bell. Under these conditions, southern state commissions played a strategic, rather than creative role. Most of their effort went directly into reducing municipal bargaining power and bringing Bell and the cities to terms quickly, generally with the result that Bell's interests predominated.

Midwestern regulators, by contrast, had to deal with their region's many independent telephone companies, who wanted to keep access to the market open and who refused to bow to Bell pressure. Years of fierce competition in the Midwest also greatly expanded telephone distribution, creating in the region a more numerous and varied clientele. These consumers, particularly those living in rural and outlying areas, demanded that state regulators consider their interests. At the same time, large, strategically important midwestern cities such as Chicago had strong traditions of public regulation. Fashioning sophisticated policies, they held on tenaciously to their regulatory authority, even with the rise of public utility commissions. With so many actors involved in policymaking in the Midwest, state commissions had to be innovative.

Poor public sector leadership also hampered public policy in the South.[159] Concerned primarily with overcoming local governmental regulation, southern state commissions conceded too much to Bell. Even when the regulatory bodies worked out compromises with the corporation, their policies ended up reflecting the same concern with limited local issues as the municipal governments they condemned. When the Georgia Railroad Commission declared telephone competition wasteful, for example, it argued in part that such competition hurt local interests who had unwisely tried to compete with Bell.[160] This paternalistic reasoning was similar to that of city officials, who used their regulatory powers to protect local capital from "foreign" corporations. The Georgia Commission ordered Bell to acquire the plant and equipment of competing firms before taking over a city, thereby "protecting" local capital.[161] Acquisition was a small price for Bell to pay for market control. The commission's policy suggests that southern regulators were satisfied with rather minor concessions from the national corporation.

Even in supposedly "progressive" states such as Georgia, politicians were not willing either to engage in regional planning that would have violated local prerogatives or to challenge Bell and gain greater concessions from the corporation. Hoke Smith, elected on an antimonopoly platform, never dared antagonize northern corporations in his state, which proved a pleasant surprise to nervous Southern Bell managers.[162] Populist outbursts about the plight of the farmer and the evils of big business divided rather than unified the South. The old ambivalence—on the one hand railing against foreign dominance, on the other bowing to outside capital in the interest of development—paralyzed the South's public sector.

As the South's inferior entrepreneurial style had caused it to miss economic opportunities early on in telephony, its weak political leadership hurt it during the era of regulation. In both cases, the preponderance of local concerns and the shortage of innovators with a broader vision were to blame. Perhaps given the South's low income, lack of urbanization, and overwhelmingly agrarian economy, little else was possible. With so much of the region's population poor, with so few of its citizens able to afford telephone service, the South had limited bargaining power with Bell. But it does seem that if politicians in the region had dropped the symbolic sideshow against "foreign" capital and eschewed periodic appeals to outmoded agrarian sentiments, which they really did not mean anyway, they might have been able to manage Bell system growth more effectively. They could have helped to spread the benefits of that system farther and wider.

In the long run, of course, regulation in the South did have one important consequence—it allowed the Bell system to grow in the face of continued local antipathy to the enterprise. Because Bell was generally unchallenged by public authorities, however, I would contend that it was less a force for development than it might have been. In the interests of profit, the corporation left large sectors of the South out of its system. On the other hand, in its attempt to create a unified network it reduced competition and eliminated local options, cutting off consumers from alternative types of service. Bell might have developed an innovative policy to bring the more remote sectors of the South more firmly into its system, as state regulation in the Midwest and Northeast forced it to do. But, concerned primarily with expanding its system in more profitable parts of the country, it too often ignored poor regions like the South and felt little pressure to change. While the Bell system helped to break down barriers of isolation and open up new opportunities for growth, change, and innovation, its revolutionary impact was muted and attenuated by the limitations of public policy in the South.

Chapter 9

Regional and Corporate Cultures

From our vantage point it is easy to conclude that the Bell system had to triumph in the South. Certainly it offered much that the people of the region lacked—rapid communications, contact with the outside world, a large-scale, technologically sophisticated industry. Yet those who participated in this history did not have the benefit of our hindsight. To southern businessmen and Bell managers alike, the growth of the Bell system was a struggle with no predetermined outcome, a battle that eventually encompassed the whole of southern society. Where one stood on the emergence of that system was a politically charged issue, which fed into debates over states' rights, econmic colonialism, and the relative merits of rural and urban cultures. The Bell system was at the center of the controversial shift in the South from the nineteenth-century world of agriculture, small towns, and rural values, to the twentieth-century one of heavy industry, large cities, and cosmopolitan society.

This basic conflict of values accompanied the emergence of large-scale business throughout the United States after 1880. As modern corporations extended their operations from the relatively advanced Northeast to the rural and small town areas that composed most of the nation, they discovered that America was not simply a flat, uninterrupted economic plane on which any edifice could be built. The United States had a traditional business culture that modern, large-scale enterprise threatened. Corporate growth severed the connection between business and society that had existed for generations.

Before the rise of big business, local firms, operating in restricted local markets, had fit well with the geographically restricted communities in which most Americans lived. The dependency of a business on its home market gave the businessman incentive to boost local development, keeping business growth and community growth in harmony. This interdependence was reinforced by a Jacksonian political economy, in which state and local governments worked to increase the opportunities available to the businessmen of each place. Business localism of this sort formed one of the strongest traditions of American capitalism; it underlay the belief that one could better one's own lot by aggressively promoting the interest of one's town or region.

By the late nineteenth century, these traditions of localism were coming into conflict with the financial, technological, and entrepreneurial demands of transregional systems like telephony. Cutting across geographical boundaries, large-scale firms challenged the well-worn relationship between business and community and undermined local control of economic activity. The result was an outcry against the corporation by members of the traditional economy. Charging monopoly exploitation, prejudicial dealing, and unfair competition, local businessmen, consumers, and farmers sought to redress these ills through their state legislatures, making use of antitrust laws and regulatory agencies to restore what they regarded as the natural order of things—small-scale, local firms.[1]

To the managers of modern corporations, such efforts to preserve local control and autonomy and make national business organizations serve local needs only interfered with efficient management. They understood that the corporation could dramatically reorganize production and distribution in ways that invariably benefited some places and hurt others, but also that their organizations could generate a new scale of wealth and material abundance. In the view of the architects of big business, therefore, challenges to the operation of their firms by local communities constituted impediments to progress. These impediments had to be removed to realize the potential of modern business enterprise. Rapid transportation, faster communications, and new technology may have driven American business to a new size and scale, but until it overcame the limitations thrown up by a locally oriented culture, it could not reach its full scope.

Beyond these sorts of problems, corporate managers also faced the enormous task of spanning the numerous individual markets that made up the locally structured American economy. Generally we think of the national market emerging effortlessly from advances in transportation and communications. In fact, large firms had to build a national market for their new products out of the existing, highly localized markets dominated by small firms. In addition, to control and coordinate the production and distribution of their mass-produced, standardized goods across the economic landscape, they had to construct organizational structures to assure the quality of output and the comparability of performance between different places. Regional variations in infrastructure, the availability or inavailability of resources, and differences in consumer tastes and political traditions made such work exceedingly difficult, particularly outside the relatively urban and affluent Northeast.

All of these elements came into play as AT&T tried to build its national telecommunications system in the South. From the very beginning, the corporation confronted conditions in the region that confounded its strategy. With the introduction of the telephone to the South, Bell managers discovered that their style of entrepreneurship differed markedly from that of southern businessmen.[2] Much like traditional businessmen elsewhere, southerners linked their own economic well-being narrowly to that of their immediate surroundings. Trying above all else to advance the value of their portfolio of interests in their communities, they had a highly local outlook. They invested in nearby projects that promoted local development, but showed little willingness to undertake the sort of system-building activity that engaged the managers of the Bell Company.

The persistence of localism in the South was very much a product of conditions in the region. It grew out of the economic situation of the South after the Civil War, when the end of slavery made investments in land and local enterprises rather than human beings the prime outlet for southern capital.[3] It was nurtured by rural poverty, isolation, and industrial underdevelopment, all of which gave southern consumers preferences for simple, inexpensive products and little use for expensive, sophisticated technological equipment. Seen from this perspective, the behavior of southern entrepreneurs made sense; it was a way of dealing with the uncertainty, narrow markets, and shortage of capital they faced. Southerners were not "precapitalist" in their outlook. They valued new industries and sought to promote regional progress, attempting to create a type of telephone service that was, as they saw it, appropriate to their society.[4] Nonetheless, this behavior led to severe problems for telephone development.

Though rational from the point of view of the individual businessman, this behavior underminded southerners' ability to respond to new opportunities. Localism made them exceedingly cautious toward new products and technologies, particularly those that went beyond their immediate surroundings or involved a large investment that could not be directly supervised. Since telephone promotion involved just such an investment strategy, southerners initially avoided it. Even when they did enter the industry, localism limited their range of action. While northern entrepreneurs began early on to lay the foundations for a regional telephone system, uniting in organizations for large-scale promotion, southerners confined themselves to small-scale, restricted endeavors, making no attempt to market the new technology on a regionwide basis. Steeped in the tradition of localism, New South businessmen focused primarily on the economic opportunities in their home market, ignoring the arguments of Bell managers that a single, integrated system was the most economical way to promote telephony. Such views gave credence to the charge that southerners on their own were incapable of fully realizing the potential of the telephone. Operating from their limited local perspective, southern businessmen could not meet the needs of an integrated national telephone network that stretched beyond the confines of any one state or region and imposed national technical standards on the whole industry.

Market forces alone could not break the logic of localism.[5] In the nineteenth-century South, a world of imperfect information, maldistributed resources, and weak markets, the resources necessary for industrial development did not move easily across geographical barriers, particularly when those barriers were reinforced by cultural differences. This problem was especially acute in industries such as telephony, which relied on complex technology and highly specialized capital goods. Bell had to work to overcome these constraints. A small group of men, drawn mainly from the North and possessing special promotional talents, provided the innovative impulse the telephone needed in the South. They transferred northern capital, technology, and business strategies to the South, while adjusting the Bell system to the region's special economic conditions. Through these efforts, they provided the sort of entrepreneurship that helped to integrate southern telephony into the national network.

James Ormes was the first of these cross-cultural entrepreneurs. Arriving in the South in the wake of the failure of indigenous southern telephone agents, he overcame southern resistance to new technology, such as the exchange. Rapidly entering major urban markets that could support this important innovation, he ignored the restricted local outlook of those who had preceded him and surveyed the entire South for opportunities. This undertaking paid off as competition for commerce among southern municipalities worked to his advantage and town after town began to demand telephone service. Ormes was able to rapidly diffuse the telephone throughout the South and prepare the region for the type of communications system Bell was beginning to develop.

Just as important as Ormes's early entrepreneurial work were his organizational innovations. Building an alliance with outside capital through Western Union, he brought needed resources to the South. With the telegraph giant's assistance, Ormes founded Southern Bell to systematically promote southern telephony. This new organization fit with the entrepreneur's early strategy by operating on a regionwide basis. Linked directly to northern and midwestern capital, staffed by outside managers, Southern Bell initially proved quite effective in building up the southern telephone industry.

Though crucial early on, Ormes's efforts failed to remove the economic, social, and political conditions of the South that gave rise to the culture of localism. Disputes flared up again after

1880, this time between Southern Bell managers and the executives at the parent company responsible for designing the Bell system. To make profits, Southern Bell had to respond to local conditions. The company's managers sought, therefore, a type of technology that fit these conditions; but their choices conflicted sharply with the technical standards set by AT&T engineers. Seeking to build an integrated telephone network, the engineers insisted that southern technology be compatible with the needs of that network.

Following the return of competition in 1894, the gulf separating the South from the emerging national network widened. New entrants rapidly brought down Bell prices and severely cut its market share by proffering new types of telephone service and equipment. In doing so, they subverted AT&T's plans to link local operations to its national toll network. Though the corporation tried to respond to the challenge, competition revealed an important flaw in its approach. AT&T had been trying to fit one system to many different economic, social, and political contexts. This policy failed to meet the needs of different regions such as the South as well as it did those of the emerging network. In the South, many poor and rural customers could not afford Bell service, which was for them unnecessarily expensive and complex. They turned instead to the cheaper alternatives offered by local telephone entrepreneurs, undercutting Bell efforts at standardization.

Gradually recognizing the flaw in its approach, AT&T eventually modified its strategy in ways that allowed it to bring the South back under control. Serving as Southern Bell president, E. J. Hall provided the innovations that made this possible. He devised a new marketing strategy that sold the Bell system to the South. Relying on his company's strength in long-distance service, he gradually expanded the network into profitable areas. Unable to compete with this high-quality service, independent firms lost these markets. Where competitors would not give way, Hall used strategic acquisitions to remove them from key points. At the same time, however, he also recognized the need to compromise. Using sublicense contracts, Hall brought many independent firms into the Bell system as separate but unequal partners. As sublicensees, local telephone companies provided service to areas that Bell could not reach, but had no chance to develop their own systems.[6]

Under Theodore Vail, AT&T applied this same strategy throughout the nation. Vail also devised a plan of functional organization that gave the parent firm strategic levers of control over its increasingly extended system, leading to a new type of Bell structure. While highly centralized in some respects, this structure also embraced several different types of organization and a broad range of technologies for the different markets it served. More flexible and responsive to regional conditions, the remodeled Bell enterprise still proved capable of bringing all areas together in a unified national system under its control.

The final, crucial element in this story was the evolution of public policy. The Bell system was simply too big, and the conflicts it generated in the South and elsewhere too deep, not to reach the political arena. Initially, municipal authorities held sway in telephone regulation, and they worked closely with local business interests to maintain a degree of local control over the industry. But just as localism was breaking down in the private sector, local political power gave way to the authority of higher levels of government. Gradually, Bell received support for its system-building activities from progressive-minded state legislators, public utility regulatory commissions, and the federal government. The corporation did not control politics, though over time Bell managers became more adept at the political game. More importantly, however, the company benefited from the congruence of its own strategies and designs with the values of certain public authorities, who rejected the tenets of localism and accepted the concept of a national system. In this way state and federal authorities generally served Bell's interests, allowing the corporation to extend its system to more and more places.

Historians studying America's transformation from a world of discreet, local communities to one of national, integrated organizations have argued that nineteenth-century society should be seen as a permeable barrier along which different local cultures interacted, not a one-way street through which eastern industrial values were imposed on recalcitrant, backward communities.[7] While there is some truth to this pluralistic view, my study reveals that in the formation of the national telephone system, Bell did have the power to impose its model of development on regions such as the South. The policies devised by Hall and Vail succeeded because to some extent they changed southern consumers' preferences for telephone service and brought them in line with company strategy. The combination of expansion and acquisition reduced southerners' options. Careful negotiations with key business elites and important political leaders in individual southern towns built support for the Bell system, turning southern boosterism to Bell's advantage. Though the formation of the modern Bell system involved compromises with regional interests in many places, all of those compromises left key levers of power—over technology, finance, and operations—in Bell's hands. And where all else failed, Bell could count on political support from sympathetic state officials willing to force the system on reluctant areas.[8]

Seen in this way, the triumph of the Bell system might be an example of corporate power run rampant. But this view is too simplistic, for the national communications network had extremely varied and complex effects on life in the South. As it penetrated the remote region, it brought both costs and benefits, neither of which was distributed uniformly across the society. Depending on where one's interests lay, the system could have been seen as the symbol of progress or the scourge of corporate monopoly. All of these factors make an accounting of the Bell system's impact on the South exceedingly difficult.

Measured strictly in terms of economic growth, the Bell system clearly brought some important benefits to the South. Intercity lines broke down small town isolation and opened up new channels through which capital and other vital resources could flow to and within the region. Entering the South through its growing towns and cities, Bell reinforced the region's most progressive and "modern" elements, located in its small urban sector.[9] In this way, the corporation seems to have created a new technological and entrepreneurial borderland between the advanced economy of the North and the backward economy of the South. Following the lines of urban growth and industrial and commercial development, this borderland helped to introduce the basic values and structures associated with large-scale business organization to the South.

The Bell telephone network also offered long-range benefits in improved communications to southern consumers that were not immediately apparent. One advantage of an interconnected network was that as it grew, it raised the utility of the telephone to all users, for it enabled them to make more connections. Like other complex technological systems, however, the telephone network was subject to problems of market failure and externalities. Individual telephone patrons in the South and elsewhere evaluated the utility of service from their individual perspective. They failed to take into account the expanded potential for making calls or being called, thereby underestimating the value of an extensive system combining local and long-distance service.[10] This sort of calculus underlay the preference of many southerners for inexpensive, rudimentary local systems. Although these preferences were perfectly rational from an individual, short-term perspective, they would change as the benefits of wider communications became apparent and the telephone evolved from a luxury item to a necessity. Committed to its vision of an integrated national network, AT&T planned service with the future in mind. The firm undertook expensive developmental work in the South in order to cultivate demand for integrated telephone service. In doing so, it helped to bring a type of technology to the region that otherwise might have taken years to arrive.

Operating from their local perspective, southern entrepreneurs were incapable of performing this work. Less constrained by local conditions, AT&T sent scarce technical, managerial, and entrepreneurial talent to the South. Motivated by the need to keep up standards and performance across its extensive system, the firm could pursue investments in telephony with important social consequences that were not in the short term profitable for southerners to undertake.

Though these benefits were real, they do not prove that the Bell system was the best choice of communications technology for the South overall. Economists, who tend to believe in the efficacy of the market for arriving at social decisions, might well question some of the results of Bell's actions. Even if there were some advantages to integrated service, did the South need as extensive a network as AT&T provided? The corporation gave the region a powerful telephone system to be sure, but it did so at the costs of higher prices and restricted distribution of telephones. How great were these costs? The answer to these questions will determine if on balance the South gained more than it lost from the Bell system.

There is no unambiguous way to resolve this important issue. In the case of business users, for example, let us assume that the savings in time was worth the cost of an extensive national telephone network. What about the rest of the population? It was possible to have a premium system that served business users who needed rapid communications over a wide area, and other systems to serve other groups of consumers, as we had before the emergence of the modern Bell system in 1907. If southern consumers truly preferred rapid long-distance voice communication over less expensive, slower forms of communication, then the South's level of investment in telephony was a wise one. Unfortunately, economics provides no method to discover consumers' "true" desires, only their revealed preferences as seen in their consumption decisions.[11]

It is too easy to assume that the eventual acceptance of the Bell system proves that the South got the sort of telephone service it wanted. With the availability of an extensive and effective telephone system, and the absence of alternatives, such as a fast, dependable postal system, people use telephones; but that does not mean that they want this arrangement above all others.[12] History shows that AT&T had to survive a strong competitive threat before its system gained wide acceptance. Had it not done so, we might now have a much different telephone system, or at least have been slower in coming to the one we do have. AT&T's success in spreading its brand of service to regions like the South even in the face of severe competition was therefore crucial in setting the industry on the path that eventually culminated in public acceptance of a national telephone network. The interpretation of that accomplishment—as meeting a public need or as establishing a monopoly—can to a large extent suggest how to view the overall value of the Bell system.

As noted above, the company's success stemmed from its ability to "market" its service in regions like the South. It performed this feat, however, not just by providing consumers with information from which to make free choices. Rather it cajoled, bribed, colluded, and at times imposed its designs on towns and cities. None of these actions made AT&T particularly reprehensible; its competitors engaged in similar tactics. If we abandon the sanitized view of the market, we can see that there is nothing inconsistent with the notion of people making choices in the face of strong pressures on their behavior. Moreover, consumers often did have the choice between Bell and alternatives. If the corporation had had to rely strictly on its coercive power, it could never have enjoyed so thorough a victory. At least some of the time it must have been giving people what they wanted.

On the other hand, this process may not have produced optimal results. When the elite members of a small southern town decided to switch their allegiance to the Bell system, they did so in the manner the economic model of rationality predicts: They calculated what was in their best interest and acted accordingly. Except that in this case their behavior had significant ramifications for the rest of their community. By choosing the Bell system, they guaranteed that later consumers, who would begin to take telephone service as their incomes and the price of that service permitted, had no choice in the matter. Once the South, and other parts of the nation, started down the path of the Bell system, there was no opportunity to change course or turn back, at least not without great cost and consternation. So while consumers' choices were rational on an individual level, we cannot be sure that they were necessarily best from a societal point of view.

This uncertainty raises a disturbing question about the process of innovation. In the absence of perfect knowledge of the future and in the presence of externalities, the market may not generate optimal choices of technology. Still, choices do have to be made. Under these conditions, the ability of a firm to strategically position itself to dominate the market for a short period may set up a path of development from which there is little possibility of change, but which is not necessarily best for society. In the case of the telephone, this uncertainty means that we cannot be sure, on the basis of consumer behavior, if the Bell system was the best choice for the South.

In the messy real world, optimal results are seldom possible, and often one is left only with the second-best alternative. Southern consumers did not have a full range of options for service from which to choose. In part, this restriction was due to AT&T's refusal to permit competition in the long-distance market and its determination to keep control of the telephone industry. But the corporation was not solely to blame. Constrained by the local entrepreneurial outlook of its business class, as well as a variety of economic conditions, the South did not have the wherewithal to construct a telephone system that both captured the long-term benefits of integrated service and provided the other characteristics that consumers desired. Thus, while the Bell system may not have been the best choice for the South, it may have been a better choice than those which would have resulted in AT&T's absence.

Granting this point, however, still leaves open a second major question. Were there other, superior methods by which the admittedly difficult and perhaps necessarily imperfect decisions about telephone technology could have been made? Although an important public issue, the choice of telephone systems in the South was largely made by private parties. Today we expect such decisions to be made with at least some political input; the breakup of the Bell system, involving a wide array of public and private interests, is an excellent example of the modern brand of policy-making.[13] In the early part of this century, however, the public mechanism for making such choices was still underdeveloped relative to private mechanisms. As a result, public policy tended only to confirm what private interests had already produced.

Most Americans are somewhat uneasy with placing so much power for important social decisions in private hands. AT&T, whatever long-term advantages it offered to the South, was still a money-making concern that based its decisions not on altruism, but on the search for profits. Its actions, however beneficial they may have been, were neither painless nor cost free. The company dominated the industry for seventeen years with a patent monopoly and fought ferociously with its competitors to maintain market share. Bell's market power extracted monopoly rents from southern consumers. The system, valuable though it may have been in the long run, offered a particular style of service that exceeded many southern consumers' needs, restricting the availability of telephones in a region where distribution was already very low. When competition decreased in the industry after 1907, expansion slowed and certain types of service declined. Rural telephony in particular suffered from Bell's monopoly position.[14]

Because of the largely private nature of telephone development, southerners for most of the period under study were caught between two worlds, as the national network emerged slowly in the face of continued southern adherence to the culture of localism. The result was a two-sector telephone industry that whipsawed many consumers between the old and the new southern economy. While parts of the South benefited from the emergence of the Bell system, others, especially rural areas, were still cut off from the advantages of modern telecommunications technology. In many cases the less advanced areas received few of the rewards, but still paid many of the costs associated with AT&T's dominance of the industry.

As a private concern, AT&T's powers to ameliorate the problems that accompanied the extension of its system to the South were quite limited. Southern economic conditions forced the company to decentralize its operations and take account of demand-side factors. This policy helped to overcome conflicts between the company and the region and allowed the Bell system to flourish in the southern environment. It also assured that, in the near term at least, corporate growth would not put an end to all vestiges of the existing southern economic and social structure. But the policy also tended to limit Bell's ability to transform the South, thereby dragging out the process of change. Only through later regulatory policies that expanded telephone service cheaply into marginal areas and used rate averaging to keep down prices did all sectors of the South begin to receive service on comparable terms.

More competition in the telephone industry might have alleviated some of these burdens. Indeed, it is even possible to imagine how a competitive market could have eventually produced an integrated national network. Instead of one dominant firm presiding over a single telephone system, we could have had multiple regional systems that eventually converged as incomes rose, prices fell, and the utility of interconnected service became manifest. This path to the national telecommunications system, however, was a costly one. With natural monopoly conditions prevailing in local service, the market could only choose one service provider at a time; thus replacing companies and technology as they became economically obsolete could only come about slowly, as municipal authorities gradually used their franchise power to shape telephone service in the face of public demands. Given the close relationship between municipal governments and local entrepreneurs, it seems quite likely that existing franchises would have become vested rights that could be wrested from their holders only after a long, exhausting legal battle.[15] This situation would have slowed down desirable innovation. If the history of American railroads offers any insight, the cost of building up then tearing down systems as they pass into obsolescence may be greater than planning an integrated system from the beginning by setting nationwide standards for service. Bell did just this through its its own operating companies and its many interconnected sublicensees.

These factors suggest that the efficacy of competition in capturing system economies is severely limited. Some greater flexibility in prices and quality of service would have been of benefit to the South, and certain adjustments of Bell technology could have been made without loss of overall network efficiency.[16] But there were limits to the degree of pluralism possible in the telephone industry if the South was to have received the benefits of membership in the Bell system. To remake the South, Bell had to treat the region as one part of its larger system, and thus subordinate southern regional needs to national goals.[17] Though costly, this approach was superior to competition as a means of bringing the long-term advantages of integrated service to the South.

Southern partisans will no doubt disagree that a Yankee corporation could have contributed to their region's progress. As southern spokesmen rightly noted, the cost of accepting assistance from northern corporations was dependency. In the case of the telephone, if the the South wanted the resources AT&T had to offer—which it clearly needed—then it had to accede to the company's concept of an integrated system. Unfortunately this sort of assistance from private organizations was the only road open to the South. A massive federal program like the New Deal might have effectively redistributed resources and promoted regional development, but in the early twentieth century no one was proposing such a policy.

State and local governments in the South were even less prepared to provide economic leadership. A misplaced fear of colonialism paralyzed the region's public sector. The power that AT&T wielded in the South made it vulnerable to the charge that it was an intruder bent on keeping southerners poor and dependent on the North. Critics of the corporation missed the point, however. It was not the presence of national corporations that was the cause of dependency, but their continued alienness within southern society.[18] Making outside business a part of southern society meant renouncing claims to a separateness of culture, giving up attempts to preserve regional "agrarian" values against encroaching industrialism, and replacing the local entrepreneurial orientation with a more cosmopolitan one.

Ironically, an acceptance of northern values would have actually given southerners more power to shape the process of development by allowing them to play a more prominent role in corporate management.[19] Had New South leaders accepted the world of big business but engaged in some hard-headed bargaining with outside corporations such as AT&T, they might have gained for their society both rapid growth and greater ability to control its impact. Rarely did this happen. Southern progressives, such as the men who staffed state public utility commissions, simply conceded to AT&T a free hand. Other political leaders rhetorically chastised "foreign" capital but, fearful of losing needed investment for their states, quickly moderated their position once the election was over and did nothing. Of course, the very weakness of the southern economy often left them little room to maneuver politically. Yet southern politicians, because they had much to gain by exploiting the fears of foreign capital and outside domination that disturbed agrarians and small-town businessmen alike, engaged more in symbolic politics than real policy-making. These practices only impoverished political discourse in the South, thereby giving greater scope to corporations such as AT&T and, in the long run, making the South even more dependent on what remained "alien" economic institutions.

While I have tried not to minimize the costs associated with AT&T's market power, the fact is that many of the problems that southerners found in the telephone industry were due as much to the limits of southern entrepreneurship, public and private, as to the activities of AT&T. Entrepreneurship is usually defined in a vague, general sort of way to mean market-oriented, profit-seeking behavior. This definition is inadequate to describe the situation in the New South. Southern businessmen aggressively pursued opportunities, but as the case of the telephone shows, they did so in ways that did not always contribute to long-run growth. Simple calculations of profit alone could not assure that southern entrepreneurs would move away from business localism and pursue innovative paths of development. Traditional patterns of business behavior could still generate healthy returns. The South was growing in this period, and within southern society the wealthy were progressing, even if relative to the rest of the nation their progress was limited. Renouncing tried and true avenues of local advancement for an unproven sort of new business activity might well have seemed unwise from an individual perspective, even if the investment might pay great future dividends for the society as a whole.

These limitations were apparent in the southern telephone industry. Both initially and again during the later years of competition, southern entrepreneurs missed opportunities presented by the telephone. Absent from the South was a large, broad-minded business class capable of taking advantage of the new opportunities the Bell system presented and diffusing them throughout the rest of the region. Cities like Birmingham and Atlanta and towns like Durham benefited from contact with AT&T; but such places were so few and far between that their gains rarely spilled over into the adjacent hinterland economy. Small-town southern businessmen, distrustful of outside capital, concerned primarily about local welfare, uninterested in the problem of rural poverty, put all of their effort into advancing the prospects of their own locality, ignoring the region's majority poor rural population.[20] While independent telephone promoters in the Midwest built their own systems, suggesting that even without Bell's assistance they would have linked individual exchanges into a network, southern telephone entrepreneurs rarely engaged in such activity. Indeed, in the Midwest, localism combined with broader system-building endeavors gave the region the best of both worlds for a time, and provided the widest distribution of telephones in the nation. In the South, by contrast, localism and entrepreneurial passivity severely restricted the distribution of telephones.

On the whole, the South gained more than it lost from the Bell system. That system may not have been the optimal choice for the region, but it was superior to possible alternatives. AT&T's method of bringing the technology of the system to the South may not have been the best one, but given the limitations of public and private actors in southern society, it was the only way for the region to realize the benefits of the national telephone network. In parts of the United States such as the Midwest, the costs of the Bell system probably outweighed its benefits. But for regions suffering from the problems that gripped the South, there was little choice but reliance on a powerful outside organization such as AT&T. Though the South did not begin to achieve parity with the rest of the nation until after World War II, the preconditions for this change were set by the sort of work done by that company. Further progress would come in the same fashion, as other outsiders—individual entrepreneurs, corporations, the federal government—crossed the invisible line separating the South from the rest of the nation and began to break down the structural conditions and prompt changes in the old habits of behavior and traditional values that held growth in check.

This conclusion suggests that large-scale corporations can be forces for progress in regional economic development. Adherents to the position that direct corporate investment in poor regions is wholly a good thing might be tempted to take the lessons of the telephone in the South to support their views. I offer two caveats for those so inclined. First, as noted above, the beneficial aspects of AT&T in the South hinged on three important factors: the inability of the market to provide the South with needed resources; the impotency of competition in the presence of externalities; and the limitations of southern entrepreneurship. Without these conditions, the value of the Bell system in the South might have been far smaller, indeed less than the costs associated with AT&T's monopoly. By the same token, there is no reason to believe that public rather than private organizations could not have provided the same benefits that AT&T did, had America's public sector been more developed.

A second warning on generalizing from the southern experience involves the nature of innovation. AT&T was a force for innovation in the South, it is true; but the history of the corporation suggests that its creativity sprang more from an early and, at the time, almost irrational commitment to the concept of an interconnected system than from any rational calculation of the costs and benefits of that system. To develop its system, moreover, AT&T had to blend political acumen with technical and business skill to carry out a difficult process of social decision-making. Perhaps rational justifications for complex decisions like this can only be made ex post facto, in the presence of perfect knowledge. I see nothing wrong with the notion that innovation, being by its very nature a process that must take place in the absence of perfect foreknowledge, has an irrational or nonrational side. But for those disposed to look for lessons and see patterns, the case of the Bell system in the South reminds us that innovation is not teleological. One should not assume that every successful innovation was successful because the innovator saw farther or looked deeper than did others into the true nature of a technology. It might just as well be true that the successful innovator proved skillful at closing off other roads and making his innovation seem like the right choice.

Notes

Introduction

1. Alan Trachtenberg, *The Incorporation of American Society: Culture and Society in the Gilded Age* (New York, 1982), 3–10.
2. Carolyn Marvin, *When Old Technologies Were New: Thinking about Communications in the Late Nineteenth Century* (New York, 1988). Marvin discusses popular reaction to these technologies in America (see esp. 191–222).
3. Robert Wiebe, *The Search for Order, 1877–1920* (New York, 1967), describes the general process of late-nineteenth century organizational development, but ignores the inability of large-scale organizations to eliminate all regional variations in economic, cultural, and political conditions.
4. John Larson, *Bonds of Enterprise: John Murray Forbes and Western Development in America's Railway Age* (Cambridge, Mass., 1984), provides an excellent analysis of these issues in railroading in the Midwest.
5. For more information on the history of the industry, see Robert W. Garnet, *The Telephone Enterprise: The Evolution of the Bell System's Horizontal Structure* (Baltimore, 1985), and George David Smith, *The Anatomy of a Business Strategy: Bell, Western Electric, and the Origins of the American Telephone Industry* (Baltimore, 1985).
6. The standard view of systems and their impact can be found in Alfred D. Chandler, Jr., *The Visible Hand: The Managerial Revolution in American Business* (Cambridge, Mass., 1977).
7. See Chandler, *The Visible Hand,* 15–75.

Chapter 1. The Telephone Heads South

1. Rosario Tosiello, "The Birth and Early Years of the Bell Telephone System, 1876–1880" (Ph.D. diss., Boston University, 1971), vi,

81–82, 89–91, 144–63. American Telephone and Telegraph Company Historical Archives (hereafter AT&T Archives), General Manager's Letterbooks (hereafter GMLB), 19, Sanders–Davis & Watts, 30 January 1879.

2. E. T. Holmes Company in Boston moved into this business, helping to form the Telephone Despatch Company. AT&T Archives, box 1116, Nathaniel Lillie Reminiscences.
3. AT&T Archives, box 1055, The Law Telephone Exchange, William Childs, 1877–78. AT&T Archives, box 1116, Nathaniel Lillie Reminiscences.
4. AT&T Archives, box 1181, New England Agency, Gower-Hubbard, 3 October 1877.
5. Information about early telephone agents can be found in my dissertation, "The Telephone in the South: A Comparative Analysis, 1877–1920" (Ph.D. diss., Johns Hopkins University, 1986), tables 3.1, 3.2.
6. Harvard University Graduate School of Business Administration, Baker Library, R. G. Dun Manuscripts (hereafter R. G. Dun), New Hampshire, vol. 16, p. 303; Maine, vol. 18, p. 84; Vermont, vol. 19, p. 205.
7. GMLB 19, Sanders–Davis & Watts, 30 January 1879. Sanders complained that "as treasurer of this company, I have no right to lend its money." Sanders also discusses the problem of capital shortages in AT&T Archives, box 1002, Letterbook of the New England Telephone Company, Sanders-Hubbard, 3 August 1878.
8. General agents may have also suffered from a shortage of funds, but if they were substantial businessmen, they had access to the resources of family, friends, and associates.
9. Garnet, *The Telephone Enterprise,* 160. The figure on southern telephones is a rough compilation gleaned from the correspondence of southern agents.
10. Richardson and Barnard had come to Savannah from Boston in 1864. They were characterized as "steady working men, doing well," in 1869. They were also said to be "well connected" with Boston capitalists and were agents for Boston steamships. R. G. Dun, Georgia, vol. 28, p. 35.
11. AT&T Archives, box 1148, Southern Agency, Richardson and Barnard, 1877–78.
12. AT&T Archives, box 1127, Early Telephone Statistics. GMLB 12, Sanders–Richardson and Barnard, 28 March 1878.

13. AT&T Archives, box 1147, Alabama Agency, Cary-Hubbard, 18 November 1877.
14. Ibid.
15. Ibid.
16. AT&T Archives, box 1178, Virginia Agency, 1877–78, Burgwyn-Hubbard, 17 December 1877.
17. Ibid., Burgwyn-Hubbard, 28 January 1878; Burgwyn-Sanders, 5 April 1878.
18. AT&T Archives, box 1178, Virginia Agency, 1877–78, Burgwyn-Sanders, 5 April 1878.
19. Ibid.
20. Ibid., Burgwyn-Hubbard, 1 March 1878.
21. Ibid., 28 January 1878.
22. Ibid., 3 March 1878; 7 June 1878.
23. GMLB 12, Sanders-Burgwyn, 26 January 1878.
24. AT&T Archives, box 1178, Virginia Agency, 1877–78, Burgwyn-Sanders, 7 June 1878. Some of his other business activities included his employment as surveyor and civil engineer for the James River Improvement Company.
25. Ibid., Burgwyn-Hubbard, 1 March 1878; 5 April 1878.
26. Ibid., 13 April 1878.
27. Tosiello, "The Birth and Early Years," 250–56, 264–68.
28. AT&T Archives, box 1178, Virginia Agency, 1877–78, Burgwyn-Hubbard, 8 January 1878; 1 April 1878; 16 March 1878.
29. Tosiello, "The Birth and Early Years," 254–56.
30. AT&T Archives, box 1178, Virginia Agency, 1877–78, Burgwyn-Hubbard, 16 March 1878.
31. Ibid., Burgwyn-Watson, 1 September 1878.
32. Ibid.
33. GMLB 12, pp. 207, 212. Garnet, *The Telephone Enterprise*, 160.
34. Garnet, *The Telephone Enterprise*, 20–24.
35. AT&T Archives, box 1178, Virginia Agency, Burgwyn-Sanders, 7 June 1878.
36. AT&T Archives, box 1148, Southern Agency, Richardson and Barnard, Richardson and Barnard–Vail, 22 March 1879.
37. Also, agents would have to keep greater stocks of equipment and instruments on hand to meet rising demand.
38. Their estimated worth was between five and twenty thousand dollars in 1877, and they had strong ties to the Boston financial community. R. G. Dun, Georgia, vol. 28, p. 35. John James, "Financial

Underdevelopment in the Postbellum South," *Journal of Interdisciplinary History* 11, 3 (1981): 443–54. James blames the structure of southern banking for inhibiting new industry in the region.

39. AT&T Archives, box 1178, Virginia Agency, Burgwyn-Hubbard, 29 March 1878.
40. Harold Woodman, "Sequel to Slavery, The New History Views the Postbellum South," *Journal of Southern History* 43, 4 (1977): 526–27.
41. Rural telephony was unprofitable and would remain so for some time.
42. Agencies were set up to encourage small entrepreneurs. Agents paid a $4.00 advance on telephones ordered and bought bells for fifty cents a piece. Given the small number of rentals in the South at this time, southern agents' inventory investments could not have been very large. Garnet, *The Telephone Enterprise,* 15–16.
43. Joseph Schumpeter, *The Theory of Economic Development* (Cambridge, Mass., 1934). For Schumpeter profits are the signal that motivates entrepreneurs; but he never explains what motivates someone to develop a new product, process, or technique before there are profits, particularly supernormal ones.
44. Lipartito, "The Telephone in the South," table 3.2.
45. GMLB 16, Vail-Cary, n.d., p. 161. AT&T Archives, box 1147, Alabama Agency, Cary-Hubbard, 4 February 1878.
46. AT&T Archives, box 1148, Alabama Agency, Richardson and Barnard–Hubbard, 29 October 1877.
47. Ibid., 20 October 1877. The agents also claimed that they would have great difficulty securing subordinates unless they could get capital from Bell to build up the business. AT&T Archives, box 1148, Georgia Telephone Development, Richardson and Barnard–Hubbard, 19 November 1877.

Chapter 2. An Injection of Entrepreneurship

1. GMLB 17, pp. 311–14, Ormes Contract.
2. AT&T Archives, box 1178, Virginia Agency, Tracy-Hubbard, 26 November 1878.
3. Ibid.
4. Southern Bell Telephone & Telegraph Company, Atlanta, Georgia, Secretary's Office (hereafter SBT&T Sect.). Correspondence of Daniel I. Carson (hereafter Carson Corresp.), Ormes–Mary Ormes,

25 September 1876. In this letter he describes his health as "improving" and states he no longer has to employ an amanuensis to do his writing. Nonetheless, his correspondence over the next ten years reveals an unsteady hand.

5. Ibid., Ormes-Hubbard (father-in-law), 31 January 1879.
6. R. G. Dun, Virginia, vol. 42, p. 295.
7. SBT&T Sect. Carson Corresp., Mary Ormes–her mother, 29 July 1873.
8. Ibid., Ormes–Mary Ormes, 25 September 1876.
9. GMLB 17, Vail-Ormes, 27 December 1878.
10. Ibid., 8 December 1878.
11. AT&T Archives, box 1148, Southern Agency, Ormes-Vail, 5 March 1879.
12. GMLB 20, Vail-Ormes, 8 March 1879.
13. SBT&T Sect. Carson Corresp., Davis and Watts–Ormes, 24 January 1879.
14. AT&T Archives, box 1264, Hubbard-Forbes, 24 May 1879.
15. AT&T Archives, box 1148, Southern Agency, Ormes-Vail, 5 March 1879.
16. Ibid., 1 March 1879.
17. GMLB 21, Vail-Ormes, 16 April 1879.
18. SBT&T Sect. Carson Corresp., Vail-Ormes, 25 February 1879.
19. GMLB 24, Vail-Ormes, 19 June 1879. Vail noted that numerous patent infringement suits would make customers wary of taking telephones. SBT&T Sect. Carson Corresp., Vail-Ormes, 25 February 1879; 26 February 1879.
20. AT&T Archives, box 1148, Southern Agency, Ormes-Vail, 14 June 1879. GMLB 24, Vail-Ormes, 19 June 1879. SBT&T Sect. Carson Corresp., Ormes-Carson, 20 June 1879.
21. The settlement with Western Union was 11 November 1880. GMLB 24, Vail-Ormes, 19 June 1879.
22. Mary Smith, "Notes on the Origins of Southern Bell," SBT&T Sect.
23. SBT&T Sect. Carson Corresp., Ormes–Carson Partnership, 8 January 1879.
24. Ibid., Ormes-Carson, 6 May 1879.
25. Ibid., 13 May 1879.
26. Ibid., Ormes-Hubbard (father-in-law), 31 January 1879; Ormes-Carson, 7 May 1879.
27. Ibid., Ormes-Carson, 7 May 1879.
28. Mary Smith, "Notes," p. 11.

29. SBT&T Sect. Carson Corresp., Ormes-Carson, 24 March 1879.
30. Ibid., Coy-Carson, 9 March 1879.
31. Ibid., Clarke-Carson, 23 April 1879.
32. Ibid., H. H. Eldred–Ormes, 12 March 1879; Maynard-Carson, 13 March 1879; Maynard-Ormes, 16 June 1879.
33. GMLB 21, Vail-Ormes, 4 April 1879; 21 May 1879.
34. Ibid., 4 April 1879. AT&T Archives, box 1109, Opening Dates.
35. SBT&T Sect. Carson Corresp., Ormes-Carson, 6 May 1879.
36. Ibid., 29 March 1879.
37. GMLB 22, Vail-Ormes, 24 April 1879.
38. Ibid., 29 April 1879.
39. GMLB 21, Vail-Ormes, 21 April 1879.
40. They demanded that Ormes pay them a 15 percent drawback on telephones he rented. AT&T Archives, box 1148, Southern Agency, Richardson and Barnard–Vail, 22 March 1879.
41. GMLB 24, Vail–Richardson and Barnard, 5 June 1879.
42. GMLB 24, Vail–Richardson and Barnard, 16 June 1879.
43. Ibid., 14 June 1879.
44. GMLB 24, Vail-Ormes, 14 June 1879.
45. AT&T Archives, box 1109, Opening Dates. The Savannah exchange was formally opened on 29 November 1879.
46. Garnet, *The Telephone Enterprise,* 61–65.
47. GMLB 20, Vail-Ormes, 8 March 1879. GMLB 23, Vail-Ormes, 20 May 1879.
48. GMLB 24, Vail–Richardson and Barnard, 5 June 1879.
49. AT&T Archives, box 1148, Southern Agency, Ormes-Vail, 23 June 1879.
50. AT&T Archives, box 1185, Stearns and George, Massachusetts, Slade–Stearns and George, 26 January 1878.
51. Lipartito, "The Telephone in the South," table 3.1.
52. C. A. Stearns and J. N. George were telegraph and electrical equipment wholesalers. In 1875, Stearns's estimated worth was $10,000 and George's was $20,000. In 1880, however, it was reported that they were extending credit to customers in their wholesale and retail businesses and were short of capital. R. G. Dun, Massachusetts, vol. 78, p. 203; vol. 86, p. 152; vol. 84, p. 83.
53. Oliver Williamson, *Markets and Hierarchies: Analysis and Antitrust Implications* (New York, 1975), 57–105. Williamson discusses the problem of proprietary interest in organizations.
54. AT&T Archives, box 1190, Western Massachusetts and Connecticut, Hayward-Gower, 6 September 1877; 18 September 1877; 15 October 1877.

55. Ibid., Hayward-Gower, 15 October 1877; 23 October 1877; 31 October 1877; 5 November 1877. In 1873 Hayward was characterized as a solid businessman said to be worth as much as $125,000. R. G. Dun, New York, vol. 324, p. 1000/LL.
56. AT&T Archives, box 1190, Western Massachusetts and Connecticut, Cheever-Bradley, 27 February 1878; Cheever-Bradley, 9 February 1878; Cheever-Sanders, 18 February 1878.
57. Ibid., Hayward–Bell Company, 13 July 1878; Hayward-Sanders, 6 April 1878; Hayward-Bradley, 10 April 1878.
58. AT&T Archives, box 1181, Rhode Island Agency, Cozzens and Bull–Vail, 24 March 1879.
59. Ibid., Cozzens and Bull–National Bell Company, 19 December 1879.
60. Ibid., Cozzens and Bull–National Bell Company, 19 December 1879.
61. AT&T Archives, box 1190, Western Massachusetts and Connecticut, Cheever-Bradley, 23 April 1878; 27 February 1878.
62. Doolittle was an inventor who, with his brother, owned a small brass-goods manufacturing company in Bridgeport, Connecticut. George Coy was part owner of a New Haven manufacturing company. In 1879 Coy took over the District Telephone and Automatic Signal Company of New Haven. Both Doolittle and Coy had strong connections to the New York capital market. R. G. Dun, Connecticut, vol. 2, p. 281; vol. 3, p. 333; vol. 41, p. 348.
63. Lipartito, "The Telephone in the South," table 3.2.
64. Richardson and Barnard demonstrated the problem in their reluctance to embrace the more expensive technology of the exchange and in their refusal to join with James Ormes in developing large urban markets. See also GMLB 24, Vail–Richardson and Barnard, 5 June 1879.
65. Of course, he was aided by Theodore Vail, who helped undermine the position of older agents and rewarded innovators such as Ormes. Garnet, *The Telephone Enterprise,* 67–69.

Chapter 3. An Organizational Solution

1. SBT&T Sect. Carson Corresp., Ormes-Carson, 7 May 1879.
2. GMLB 20, Vail-Ormes, 24 February 1879. GMLB 23, Vail-Hinkley, 27 May 1879. GMLB 33, Madden-Noble, 30 December 1879; 31 December 1879. AT&T Archives, box 1158, Davis and Watts, Agents in Maryland. The manufacturers listed the cost of an office

bell at $2.50, a battery at $2.50, and annunciator drops at $1.50 each.

3. SBT&T Sect. Carson Corresp., Coy-Carson, n.d.
4. Ibid., Ormes-Carson, 29 March 1879.
5. Ibid.
6. Ibid., Munson-Ormes, 28 May 1879.
7. Ibid., J. O. Jeffries–Ormes, 1 June 1879.
8. Ibid., Vail-Ormes, 31 March 1879.
9. R. G. Dun, Georgia, vol. 13, p. 418.
10. SBT&T Sect. Carson Corresp., Ormes-Carson, 7 May 1879.
11. GMLB 22, Vail-Ormes, 29 April 1879.
12. AT&T Archives, box 1034, Atlanta Exchange Partnership, J. J. Storrow–Vail, 9 January 1880.
13. Ibid., Ormes-Vail, 31 October 1879; 10 November 1879; J. J. Storrow–Vail, 9 January 1880.
14. Ibid., Atlanta Exchange–National Bell, 9 December 1879.
15. SBT&T Sect. Carson Corresp., Ormes-Carson, 15 July 1879.
16. Kenneth Stampp, *The Era of Reconstruction, 1865–1877* (New York, 1965), 161.
17. R. G. Dun, Georgia, vol. 7, pp. 261, 308.
18. AT&T Archives, box 1034, Atlanta Partnership, Brown-Vail, 20 November 1879; Atlanta Partnership–National Bell, 17 December 1879.
19. Ibid., Brown-Vail, 20 November 1879.
20. Ibid., Ormes-Vail, 10 November 1879.
21. Physical connection between exchanges was only possible over limited distances at this time, but in the ensuing decade long-distance communications would become an important part of the telephone business. Ormes may have foreseen this development, as Theodore Vail apparently did, and have been preparing the South for it through his strategy.
22. AT&T Archives, box 1034, Atlanta Exchange Partnership, Atlanta Partnership–National Bell, 9 December 1879; Ormes-Vail, 31 October 1879.
23. Ibid., Brown-Vail, 20 November 1879; 17 December 1879.
24. Ibid., Ormes-Vail, 31 October 1879.
25. Ibid., 31 October 1879; 10 November 1879. Not all members of the Bell organization felt this was the right move. Attorney J. J. Storrow believed that Bell still needed to rely on local support, and therefore advised a compromise between Ormes and his partners. Storrow-Vail, 9 January 1880.

26. The Atlanta episode took place largely after Ormes had already established contact with outside interests, which itself may have increased the Browns' fears and led to the problems. But conflict in Atlanta had been brewing for months and may have become open only after Ormes made his intentions clear. SBT&T Sect. Carson Corresp., Ormes-Carson, 21 August 1879.
27. David Carlton, *Mill and Town in South Carolina, 1880–1920* (Baton Rouge, La., 1982), 40–81. See also Gavin Wright, *Old South, New South: Revolutions in the Southern Economy since the Civil War* (New York, 1986), 20–24.
28. Lance Davis, "Capital Immobilities and Finance Capitalism: A Study of Economic Evolution in the United States, 1820–1920," *Explorations in Entrepreneurial History,* 2d series, 1, 1 (1963): 89–91.
29. Partly this was due to the "opportunistic" nature of southern entrepreneurship and its lack of a more sophisticated "developmental" component. Or more precisely, in the South, "developmental" entrepreneurship was aimed mainly at building up the local economy. See Maury Klein and K. Yamamura, "The Growth Strategies of Southern Railroads," *Business History Review* 41, 4 (1967): 358–77.
30. SBT&T Sect. Carson Corresp., Ormes-Carson, 6 May 1879.
31. GMLB 21, Vail–Davis and Watts, 21 May 1879.
32. AT&T Archives, box 1148, Southern Agency, Ormes-Vail, 14 June 1879. Lance Davis, "Capital Immobilities and Finance Capitalism," notes that manufacturers often supported other new industrial enterprises. Baltimore was also a traditional source of capital for the South.
33. GMLB 21, Vail-Ormes, 21 April 1879. GMLB 24, Watson-Ormes, 14 June 1879. Vail complained that "every agent thinks his territory is the keystone and will not listen to any excuses for delays." GMLB 17, Vail-Williams, 21 December 1878. Tosiello, "The Birth and Early Years," 143–44, 149–50.
34. GMLB 17, Vail-Williams, 21 December 1878. Tosiello, "The Birth and Early Years," 141–42, 145, 158–63.
35. SBT&T Sect. Carson Corresp., Jeffries-Carson, 17 June 1879.
36. Ibid., Ormes-Carson, 24 May 1879.
37. GMLB 22, Vail–Davis and Watts, 9 May 1879. GMLB 14, Vail–Davis and Watts, 31 May 1878. Ormes already relied on them to some extent. SBT&T Sect. Carson Corresp., Davis-Ormes, 22 March 1879.

38. GMLB 21, Vail–Davis and Watts, 28 May 1879. GMLB 24, Vail–Davis and Watts, 11 June 1879. AT&T Archives, box 1033, J. D. Martin, Early Telephone History.
39. SBT&T Sect. Carson Corresp., Davis and Watts–Ormes, 24 January 1879; Ormes-Carson, 15 May 1879; 15 July 1879. The manufacturer's business was secure but small, reportedly worth seventy-five to one hundred thousand dollars. At the time Ormes needed them they were short of funds, overextended in serving their rather far-flung market. R. G. Dun, Maryland, vol. 13, p. 4; vol. 15, p. 116.
40. GMLB 24, Vail–Davis and Watts, 11 June 1879. GMLB 25, Vail-Ormes, 27 June 1879.
41. GMLB 21, Vail–Davis and Watts, 21 May 1879. GMLB 18, Sanders–Davis and Watts, 18 January 1879. GMLB 25, Vail–Davis and Watts, 27 June 1879. Smith, *The Anatomy of a Business Strategy,* 86–88. Smith discusses Vail's reasons for not decentralizing telephone production.
42. He had also gotten equipment for Atlanta from Post and Company in Cincinnati when Davis and Watts were too slow. SBT&T Sect. Carson Corresp., Ormes-Carson, 16 July 1879; Electrical Merchandising Company–Ormes, 5 May 1879; Ormes-Carson, 11 July 1879; Davis and Watts–Ormes, 29 March 1879.
43. Smith, "Notes," Biographical Information.
44. SBT&T Sect. Carson Corresp., Ormes-Carson, 24 May 1879.
45. AT&T Archives, Contract Book.
46. SBT&T Sect. Carson Corresp., Vail-Ormes, 2 June 1879.
47. Smith, *Anatomy,* 76–79.
48. SBT&T Sect. Carson Corresp., Ormes-Carson, 23 June 1879; Vail-Ormes, 24 June 1879; Ormes-Carson, 1 July 1879.
49. Smith, *Anatomy,* 77.
50. SBT&T Sect. Carson Corresp., Ormes-Carson, 10 July 1879; 11 July 1879.
51. Ibid., 11 July 1879.
52. Ibid. Emphasis in original.
53. Ibid., 1 August 1879. Emphasis in original.
54. Ibid., 8 August 1879; 12 August 1879.
55. Ibid., 19 August 1879. Emphasis in original.
56. Ibid.
57. Ibid., 15 August 1879.
58. Ibid., 29 August 1879.
59. SBT&T General Ledgers, vol. 1 (hereafter GL 1), p. 383. The commissions paid to Southern Bell by Western Union only

amounted to about $120 per month between 1881 and 1882. Southern Bell had to pay a royalty of almost $3,000 per month to Western Union.

60. Garnet, *The Telephone Enterprise,* 44–55.
61. SBT&T Sect. Carson Corresp., Ormes-Carson, 8 August 1879.
62. SBT&T *Annual Report,* 1881, p. 1 (city: number of subscribers): Richmond: 270; Norfolk: 184; Peterborough: 82; Augusta: 88; Charleston: 125; Savannah: 139; Danville: 32; Lynchburg: 77; Wilmington: 106; Mobile: 93; Raleigh: 50.
63. SBT&T *Annual Report,* 1881, p. 1.
64. SBT&T Stock Book, 1880.
65. Ibid. Richardson and Barnard bought 82 additional shares, the full extent of their involvement in Southern Bell. SBT&T Sect., Minute Books (hereafter SBM), 19 October 1881.
66. Not all trading was in new stock because Ormes sold some of his franchise shares immediately, probably to reap a windfall profit and to pay back some of his early silent backers. Other early investors also bought and sold stock within the first year.
67. SBT&T Sect., Mary Smith, "List of Stockholders."
68. SBT&T Corporate History, "Stock Issued and Dividends Paid, 1880–1920."
69. Ibid. See also SBT&T Corporate History, "Stock Digest," and SBT&T Stock Book.
70. SBT&T Stock Book. John Munson, a speculator in Southern Bell stock, was an investor who specialized in western farm mortgages. By 1889 his letterhead also listed him as a specialist in telephone stock.
71. The company gave its suppliers free telephone service as payment whenever it could. SBM, 27 May 1880; 15 July 1880.
72. Thomas Navin and Marian Sears, "The Rise of a Market for Industrial Securities," *Business History Review* 29, 2 (1955): 105–38.
73. AT&T Archives, box 1034, Southern Bell Organization.
74. AT&T Archives, box 1210, Acquisition of Interest, Vail-Forbes, 26 December 1882.
75. SBT&T Stock Book. SBT&T Corporate History, "Historical Data."
76. SBT&T Corporate History, "Stock Issued and Dividends Paid, 1880–1920."
77. Clark Wilson, *Tracing the Telephone in Western Massachusetts* (Springfield, Mass., 1958).
78. AT&T Archives, box 1185, Lowell District Telephone Company, Longley Memorandum, pp. 3–5.
79. AT&T Archives, box 1016, F. V. Storey, "New England Telephone

and Telegraph Company Corporate History," 11–13. See Garnet, *The Telephone Enterprise,* 18–20, for a discussion of the New England Bell Company.

80. AT&T Archives, box 1185, Lowell District Telephone Company, 1878–80, Glidden-Bradley, 25 March 1879.
81. Ibid., Glidden-Vail, 29 May 1879.
82. AT&T Archives, box 1125, Charles Glidden Biography. AT&T Archives, box 1213, Licensee Company Organizational Statistics, 1882–83.
83. Wilson, *Tracing,* 99–106.
84. AT&T Archives, box 1074, Joel C. Clark, Clark-Vail, 12 February 1884. Clark managed to merge country and suburban telephone exchanges into the Central Massachusetts Company on 25 November 1879. See also AT&T Archives, box 1015, Central Massachusetts Telephone Company, 1880–81.
85. AT&T Archives, box 1015, Central Massachusetts Telephone Company, 1880–81, Memorandum, n.d. The Lowell group, though heavily involved in regional promotion in New England, acted like a modern business organization, seeking profits wherever they arose, not tied to any one location. The syndicate, for example, branched out by acquiring the Northwest Telephone Company of Minnesota.
86. Ibid., Cutler-ABT, 25 May 1880.
87. AT&T Archives, box 1015, Central Massachusetts Telephone Company, 1880–81, Memorandum, n.d.
88. Ibid. This favorable view of consolidation enabled Clark to find a place at Bell as a manager.
89. Not to be confused with the earlier, unsuccessful New England Bell Company.
90. Most of the firms and individuals carrying out consolidation were Bell licensees. That is, they operated under a franchise to lease Bell telephones. NET&T was a Bell-licensed company, but it was owned directly by American Bell.
91. AT&T Archives, box 1116, Nathaniel Lillie Reminiscences. Through these new organizations, Bell was able to take greater control of entrepreneurial and managerial decisions in the industry. It also gained greater influence over regional firms by buying their stock, thus receiving profits directly from their work.
92. AT&T Archives, box 1125, Charles Glidden Telephone Activities. AT&T Archives, box 1116, Nathaniel Lillie Reminiscences. Clark, *Tracing,* 152–53. Glidden received a managerial position at Bell.
93. AT&T Archives, box 1190, Western Massachusetts and Connecticut, Cheever-Bradley, 27 February 1878. Cheever complained that

a contract provision which allowed Bell to buy out their exchange for 15 percent profit was unfair because the investors had undertaken considerable risk and were entitled to whatever returns the business would bring once it got going. The costs associated with entering a new business—learning to operate it efficiently and taking early entrepreneurial risks—may all be compensated by the expectation of a higher return later. This is what early agents had in mind and explains why they frequently disputed Bell over the terms of their contracts.

94. In small midwestern towns, where resistance to change was strong, Western Union also provided capital and entrepreneurial leadership to induce change. AT&T Archives, M. D. Attwater, "The History of the Central Union Telephone Company" (26 August 1913).
95. Rondo Cameron, *Banking and Economic Development* (New York, 1972), 10–11. Cameron argues that large organizations help to compensate for scarce entrepreneurial talent.
96. AT&T Archives, Agreements, book 1. Connecticut and Rhode Island were run by the Southern New England Telephone Company, a firm in which Bell maintained only a minority interest.
97. The firm had the monopoly power necessary to sweep aside indigenous constraints on telephone growth. The role of monopoly in innovation is discussed by Schumpeter in *The Theory of Economic Development.*
98. SBM, 1884 Stock Issue.
99. Smith, "Notes," Biographical Information. Ormes also received a franchise for seventeen small towns in the Midwest, as well as one for the rural counties in the Philadelphia - New York area. Perhaps Bell managers as well recognized that he had developed special skills for promoting modern technology in unpromising markets. AT&T Archives, box 1178, *Southern Herald of the Telephone,* Ormes obituary. Ormes died in Switzerland in 1895.
100. SBT&T Sect., James Ormes, "Venetian Sketches and Glimpses," 1886.
101. AT&T Archives, box 1265, National Company Policy, Vail, 28 March 1881.

Chapter 4. Rapid Change and Technological Conflict

1. AT&T Archives, box 1011, Building Early Long Distance Lines, Doolittle-Hall, 23 July 1885; Hall-Vail, 12 May 1885.
2. SBT&T *Annual Report,* 1883. The report stated that reconstruc-

tion costs were high because wires had to be removed from rooftops and put on permanent poles. The company also had to lay out money to acquire the Atlanta exchange and the private-line business from James Ormes. Construction costs from 1882 to 1885 were as follows: 1882—$74,397; 1883—$46,263; 1884—$72,309; 1885—$41,542. Construction costs per customer had declined significantly between 1880 and 1885, as Southern Bell added almost 2,900 new subscribers.

3. Thomas P. Hughes, "Inventors: The Problems They Choose, the Ideas They Have, and the Inventions They Make," in P. Kelly and M. Kranzberg, eds., *Technological Invention* (San Francisco, 1978). Hughes discusses the relationship between inventors' choice of problems to solve and the business environment. See also Thomas P. Hughes, *Networks of Power: Electrification in Western Society* (Baltimore, 1983). He employs the idea of technological style to explain how economic, political, and cultural conditions affect technology. Garnet, *The Telephone Enterprise,* 69–83, describes some of the problems to which northern Bell engineers devoted their efforts.
4. SBT&T, Corporate History, "Page Value of Bonds Issued." SBT&T *Annual Reports,* 1880–89.
5. SBT&T, Corporate History, "Stock Issued and Dividends Paid, 1880–1920."
6. Southern Bell Revenues, 1882–87

Year	% Change Exchange Revenue	% Change Private-Line Revenue
1882–83	38	35
1883–84	36	65
1884–85	16	−6.6
1885–86	13	−3.1
1886–87	11	9.3

Source: SBT&T *Annual Reports,* 1880–90.

7. Garnet, *The Telephone Enterprise,* 77–79. Garnet notes that Bell had to subsidize these operating company expenditures.
8. The telegraphic equipment had been used in lawyers' offices in New York, hence the name. AT&T Archives, box 1055, Law Telephone System, William Childs, 1877–88.
9. Louderback rated the Law board highly. SBT&T Sect. Carson Corresp., Louderback-Carson, 12 December 1879; Ormes-Carson, 2 December 1879.
10. AT&T Archives, box 1055, Law System v. Multiple Board,

Lockwood-Hudson, 11 April 1885. The equipment was in use in St. Louis, Oswego (New York), Philadelphia, as well as Atlanta, Richmond, and Norfolk.

11. AT&T Archives, box 1236, Law System Advantages and Disadvantages, Lockwood-Hudson, 14 October 1887. Lockwood noted that among its advantages were its speed, simplicity, and low cost of apparatus and maintenance. Efficiency is a more debatable point, but most of the board's problems in this area resulted only when capacity had to be expanded significantly.
12. Ibid. AT&T Archives, box 1144, Telephony, Early State of the Art, 37–38.
13. AT&T Archives, box 1236, Law System, Advantages and Disadvantages, Lockwood-Hudson, 14 October 1887.
14. Ibid., Hayes-Hudson, 15 October 1887; AT&T Archives, box 1055, Law System v. Multiple Board, Lockwood-Hudson, 11 April 1885. Multiple switchboards are more than one board in a central office, each having the capacity to make all possible subscriber connections. They were necessary in high-density areas because one operator could not handle all incoming calls. F. L. Rhodes, *Beginnings of Telephony* (New York, 1929), 54–56.
15. AT&T Archives, box 1225, Bell Circular 1884. Daniel Carson noted the lower density of subscribership in the South, and the feasibility of using the Law system. In 1889, for example, the average southern exchange had 155 customers. At roughly the same time (1887), exchanges in New England averaged 184.3. See SBT&T *Annual Report,* 1889. AT&T Archives, box 1245, *Annual Reports,* 1884–89. On the other hand, those in the New York and New Jersey Company territory, which included northern New Jersey and Long Island, averaged 146. Clearly, however, the potential number of subscribers was significantly greater in the North than in the South at this time.
16. AT&T Archives, box 1236, Law System, Use by East Tennessee Telephone Company, Wilson-Phillips, 17 October 1887.
17. Ibid., Phillips-Stockton, 24 October 1887.
18. AT&T Archives, box 1076, Telephone Exchange System, 1892–93, Hayes-Hudson, 16 January 1893.
19. AT&T Archives, box 1236, Law System, Use by East Tennessee Telephone Company, Brown-ABT, 6 May 1887; Barton-Hudson, 28 April 1887. AT&T Archives, box 1236, Mann (Law) System, Doolittle-Hudson, 22 September 1891. AT&T Archives, box 1236, Western Electric Patents, Barton-Hudson, 31 May 1888; 8 June

1888. AT&T Archives, box 1055, Law System v. Multiple Board, Lockwood-Hudson, 11 April 1885. AT&T Archives, box 1076, Telephone Exchange Development, 1892–93, Hayes-Hudson, 16 January 1893. AT&T Archives, box 1236, Law System, Advantages and Disadvantages, Lockwood-Hudson, 14 October 1887.

20. AT&T Archives, box 1236, Western Electric Patents, Barton-Hudson, 31 May 1888.
21. Garnet, *The Telephone Enterprise,* 70–72.
22. AT&T Archives, box 1240, Switchboard, Infringement of Western Electric Patents, Lockwood-Hudson, 16 August 1893.
23. Earlier Theodore Vail had warned operating companies of the danger of reliance on outside equipment. AT&T Archives, box 1055, Law System v. Multiple Board, Vail-Lockwood, 11 April 1885.
24. AT&T Archives, box 1236, Western Electric Patents, Barton-Hudson, 6 June 1888; 8 June 1888.
25. AT&T Archives, box 1240, Switchboard, Infringement of Western Electric Patents, Hayes-Davis, 11 July 1895; Hayes-French, 14 June 1895.
26. Reese Jenkins, *Images and Enterprise: Technology and the American Photographic Industry, 1839–1925* (Baltimore, 1975). Jenkins develops the idea of a "business-technological mindset" to explain how technological innovation affects the creation of new industries and the development of new products.
27. AT&T Archives, box 1258, Charles McCluer Patents, Royalties.
28. Ibid., Swann-Hudson, 23 February 1895.
29. Ibid., McCluer-Hudson, 2 May 1890.
30. AT&T Archives, box 1253, Charles McCluer, Automatic Exchange System, 1893–94.
31. SBM, 8 May 1889.
32. AT&T Archives, box 1258, Charles McCluer Patents, Royalties, Lockwood-Hudson, 29 August 1889.
33. Ibid., Hayes-Hudson, 1 July 1889.
34. Ibid., Lockwood-Hudson, 14 May 1890; Lockwood-Hudson, 21 October 1890.
35. This idea of innovation is developed more fully in Hughes, "Inventors."
36. Garnet, *The Telephone Enterprise,* 74.
37. AT&T Archives, National Telephone Exchange Managers Association Meeting (hereafter AT&T NEA Meeting) 1881, pp. 140–41, 184–85.
38. Ibid., pp. 120–41. AT&T Archives, box 1223, Mechanical Department Reports, 1885–87, Letter to Vail, 14 April 1885.

39. AT&T NEA Meeting, 1881, pp. 120–41.
40. Zvi Griliches, "Hybrid Corn and the Economics of Innovation," in Robert Fogel and S. Engerman, eds., *The Reinterpretation of American Economic History* (New York, 1971). Griliches discusses the effects of demand on the diffusion of new technology.
41. AT&T NEA Meeting, 1885, pp. 32–48.
42. AT&T Archives, box 1240, Thomas Lockwood, Miscellaneous Correspondence, Lockwood-Hudson, 15 April 1885.
43. AT&T NEA Meeting, 1889, John J. Carty, F. E. Pickernell, and A. S. Hibbard, "The New Era in Telephony."
44. AT&T NEA Meeting, 1889, p. 58.
45. AT&T Archives, box 1263, Southern Managers Meeting.
46. AT&T NEA Meeting, 1890, E. J. Hall, "Corporate Organization."
47. AT&T Archives, box 1263, Southern Managers Meeting.
48. Ibid., p. 39.
49. This was a major concern of Lockwood's early on. See AT&T Archives, box 1240, Thomas Lockwood, Miscellaneous Correspondence, Lockwood-Hudson, 17 March 1886. Leonard Reich, "Industrial Research and the Pursuit of Corporate Security: The Early Years of Bell Labs," *Business History Review* 54, 4 (1980): 504–29. Reich discusses the notions of offensive and defensive industrial research.
50. AT&T Archives, box 1315, Standardization of Apparatus, Davis-Hudson, 18 April 1899. AT&T Archives, box 1035, Engineering Department, Organization and Function, 1894–99, Davis-Hudson, 1 January 1894; 28 March 1899.
51. AT&T Archives, box 1315, Standardization of Apparatus, Davis-Hudson, 18 April 1899; Hayes-Davis, 11 April 1899. AT&T Archives, box 1282, Equipment Quality Control, Western Electric, 1895, Hayes-French, 4 April 1895.
52. Reich, "Industrial Research," notes that before 1902 AT&T research was dedicated to protecting existing technology and regulating the internal structure of the organization through patent acquisition and routine engineering. Only when control of technical departments passed from Hammond Hayes to J. J. Carty did theoretical research become a major concern. Neil Wasserman, *From Invention to Innovation: Long-Distance Telephone Transmission at the Turn of the Century* (Baltimore, 1985), has a different view.
53. Garnet, *The Telephone Enterprise*, 112–24.
54. AT&T Archives, box 1011, Building Early Long Distance Lines, Hall-Hudson, 21 January 1888.
55. Allen Pred, "The Growth and Development of Systems of Cities in

Advanced Economies," in Allen Pred and G. E. Tornqvist, *Systems of Cities and Information Flows: Two Essays,* Lund Studies in Geography, series B, 38 (1973): 1–82. Pred discusses the impact of such factors on innovation.

56. This point is discussed in later chapters.
57. AT&T Archives, box 1113, Pan Electric Organization, 1883–84. AT&T Archives, box 1225, J. Harris Rogers Application for Employment, 1884, Rogers-Vail, 6 March 1884.
58. SBT&T, *Annual Report,* 1887.
59. GMLB 198, Tompkins-Glidden, 5 August 1885. Long had organized the Pan Electric Company, but left to found his own firm after a dispute.
60. John Brooks, *Telephone: The First Hundred Years* (New York, 1976), 76–81.
61. AT&T Archives, box 1239, Thomas Perrin, Multiple Switchboard Patents, Lockwood-Hudson, 7 October 1891.
62. AT&T Archives, box 1113, Pan Electric Organization, 1883–85, Rogers-Vail, 3 May 1883.
63. AT&T Archives, box 1239, Thomas Perrin, Multiple Switchboard Patents. AT&T Archives, box 1225, J. Harris Rogers Patents, 1883–84, Lockwood Report.
64. AT&T Archives, box 1225, J. Harris Rogers Patents, 1883–84. AT&T Archives, box 1113, Pan Electric Organization, 1883–85, Lockwood Memorandum.
65. AT&T Archives, box 1119, Publication of Hostile Pamphlets, Charles Swann, 1878–87. Robert Bruce, *Bell: Alexander Graham Bell and the Conquest of Solitude* (Boston, 1973), 275–76. AT&T Archives, box 1225, Pan Electric Activities, Circular of Pan Electric Company of Alabama, 21 October 1884.
66. Brooks, *Telephone,* 88–89.
67. This would become readily apparent after Bell's patents expired in 1894. See chap. 5.
68. For a good discussion from the parent company's point of view see Garnet, *The Telephone Enterprise,* 96–103. See also AT&T Archives, box 1240, Thomas Lockwood, Miscellaneous Correspondence, Lockwood-Hudson, 17 March 1886; 12 January 1890.

Chapter 5. Market Challenges

1. Gerald W. Brock, *The Telecommunications Industry: The Dynamics of Market Structure* (Cambridge, Mass., 1981), 110–14, 124.

2. John V. Langdale, "The Growth of Long-Distance Telephony in the Bell System, 1875–1907," *Journal of Historical Geography* 4, 2 (1978): 150–52.
3. For a discussion of the economics of quality choice, see A. Michael Spence, "Monopoly, Quality, and Regulation," *Bell Journal of Economics* 6, 2 (1975): 417–29. See also Brock, *The Telecommunications Industry*, 107.
4. AT&T Archives, box 1259, Long Distance Service, Extension of, 1894, Doolittle-Hudson, 8 March 1894; 9 March 1894; 10 March 1894.
5. AT&T Archives, box 1011, Building Early Long Distance Lines, E. J. Hall Memo, 24 September 1887.
6. Indeed, the definition had been formed to judge the quality of long-distance communications.
7. *Huntsville Mercury*, 31 January 1900.
8. AT&T Archives, box 66, Sub-Licensing Policy, 1907–8, Hall-French, 16 May 1908.
9. AT&T Archives, box 1011, Building Early Long Distance Lines, Hall Memo, 24 September 1887.
10. Here and throughout, "local" is used to mean service within an exchange area, generally a two-mile radius of an exchange.
11. Philadelphia was the one major eastern city where competitors made a significant challenge.
12. Compare Richard Gabel, "The Early Competitive Era in Telephone Communication, 1893–1920," *Law and Contemporary Problems* 34 (Spring 1969): 340–59.
13. AT&T Archives, box 1263, Alabama Telephone and Construction Company, 1896, Carson-ABT, 15 April 1896; Sharp-Easterlin, 16 April 1896.
14. Ibid., interview with George Wilkins.
15. As Southern Bell agent A. G. Sharp put it, "I wanted the thing closed and closed quick, so [I] paved the way to cut it off short." Ibid., Sharp-Easterlin, 16 April 1896. See also SBM, 22 April 1896.
16. GMLB 631, Sharp-French, 26 August 1901.
17. Ibid.
18. Ibid.
19. Ibid.
20. Ibid.
21. The Viaduct Company, ironically, was the successor to Davis & Watts of Baltimore, the firm to which James Ormes had earlier turned for support.

22. GMLB 631, Sharp-French, 26 August 1901.
23. Ibid.
24. Between 1880 and 1900 per capita income in North Carolina grew by 17 percent, about the average for the rest of the seaboard South. On the other hand, the state's growth was faster than Georgia's, South Carolina's, and Alabama's. In these latter three states, nonindustrial output per worker declined precipitously, while it remained stable in North Carolina. Virginia and Florida grew even faster than North Carolina and also experienced large increases in independent telephones. Richard Easterlin, "Interregional Differences in Per Capita Income, Population, and Total Income, 1840–1950," tables A–1, A–2, in Conference on Research in Income and Wealth, *Trends in the American Economy in the Nineteenth Century* (Princeton, 1960).
25. AT&T Archives, box 1163, North Carolina Interstate Telephone Company, 1900, Map of Lines.
26. Ibid., Prospectus.
27. Ibid.
28. Ibid.
29. Ibid. This does not necessarily mean that the company was using the latest equipment, only that the equipment that it had was of good quality.
30. Ibid., Map of Lines.
31. U.S. Bureau of the Census, *Telephones and Telegraphs and Municipal Fire Alarm and Police-Patrol Systems, 1912* (Washington, D.C., 1915), table 30, p. 41. The difference between Bell and independent operations was by far the most striking in North Carolina.
32. AT&T Archives, box 1348, Sub-licensees, Advantages to Operating Companies, 1903, Hall-Fish, 31 July 1903.
33. AT&T Archives, box 2019, Southern Bell Progress Report, 1911.
34. Florida's per capita income grew 169 percent in the years 1880–1900. Its agricultural income per worker shot up almost 22 percent, the highest in the region. Easterlin, "Interregional Differences in Per Capita Income, 1840–1950," tables A–1, A–2.
35. Atlanta Historical Society, Atlanta, Georgia. West Palm Beach Telephone Company History, folder 9.
36. Ibid.
37. Ibid.
38. AT&T Archives, box 1214, Rural Telephone Service, 1899–1902. AT&T Archives, box 1342, Rural Telephone Service, 1903–4. AT&T Archives, box 1279, Exchange Radii and Branch Lines, Def-

inition of, 1896. Often Bell made farmers build their own lines to the nearest exchange. Even when farmers could use an existing Bell toll line for this connection, they were charged toll rates for the service. Many farmers objected to this "extra" charge, noting that town residents paid only a fixed fee for similar service. The only way for Bell to eliminate these costs was to further extend its system until it touched even remote rural places. But expansion took time, and few farmers were willing to wait.

39. AT&T Archives, box 1372, Rural Telephone Service, 1905–6, Hayes Report. AT&T Archives, box 1357, Chicago Telephone Company, Operating Organization, Memo, 1909, Carty-Thayer, 1 March 1909.
40. AT&T Archives, box 1363, Rural Telephone Service, 1907–10.
41. The Populists made nationalization of the telephone industry part of their 1896 party platform. This issue was first raised in the Omaha Platform of 1892. George Tindall, *A Populist Reader* (New York, 1966), 94. See also Claude Fisher, "The Revolution in Rural Telephony, 1900–1920," *Journal of Social History* 21, 1 (1987): 5–26.
42. Brock, *The Telecommunications Industry,* 107, 111.
43. AT&T Archives, box 1372, Rural Telephone Service, 1905–6, Pickernell-Fish, 6 December 1906. See also box 1363, Southern Bell Rural Development.
44. United States Bureau of the Census, *Telephones and Telegraphs, 1902* (Washington, D.C., 1906), 34–35.
45. Places with 4,000 people include some fairly substantial southern towns. This may also distort the figures.
46. SBT&T, *Annual Report,* 1900–1902. By 1902 there were a total of 9,600 rural telephones in the South. Bureau of the Census, *Telephones and Telegraphs, 1902,* table 38.
47. AT&T Archives, box 1372, Rural Telephone Service, 1905–6, Pickernell-Fish, 6 December 1906.
48. Garnet, *The Telephone Enterprise,* 103–4. See also Harry B. McMeal, *The Story of Independent Telephony* (Chicago, 1934). AT&T Archives, box 1039, Central Union Financing and Reorganization, 1883–1912. AT&T Archives, box 1033, Central Union Organization and Development, 1883–1912, Allen-Fish, 11 February 1903.
49. AT&T Archives, box 1116, William Allen, History of the Independent Companies.
50. Bell exchanges in the Midwest were on the average larger than those of independents, suggesting that Bell controlled the larger towns and cities. Figures below are from ibid.

Bell and Independent Customers per Exchange, 1902

	Bell Ex.	Cust./Ex.	Ind. Ex.	Cust./Ex.
South	108	380	458	121
New England	473	324	126	100
Midwest	679	472	2,832	162
Far West	242	295	550	114
Pacific	596	258	38	125
Mid Atlantic	946	396	785	177

Note: Bell Ex. = number of Bell exchanges; Ind. Ex. = number of independent exchanges; Cust./Ex. = customers per exchange.

51. AT&T Archives, box 1033, Central Union Organization and Development, 1883–1912, Allen-Fish, 11 February 1903.
52. Ibid. See also Gabriel Kolko, *The Triumph of Conservatism: A Reinterpretation of American History, 1900–1916* (New York, 1963), 47–49.
53. AT&T Archives, box 1033, Central Union Organization and Development, 1883–1912, Allen-Fish, 11 February 1903. Also, AT&T Archives, box 1337, Central Union Company, Report on, 1902–4. There is evidence of conflict between American Bell and operating company management. The parent organization's orientation to high-quality service and system integration made it insensitive to the needs of companies serving hinterland regions.
54. AT&T Archives, Toll Maps. The maps show that AT&T's system grew by connecting major cities along heavy use routes first, then by filling in the gaps. It was in these gaps that the independents could build their own long-distance network. See also Ronald Abler, "The Telephone and the Evolution of the American Metropolitan System," in Ithiel de Sola Pool, ed., *The Social Impact of the Telephone* (Cambridge, Mass., 1977). Also, McMeal, *Independent Telephony,* 81–84.
55. James B. Hoge, "National Inter-State Telephone Association," *The Telephone Magazine* (July 1905): 34.
56. AT&T Archives, box 1337, Interstate Independent Telephone Association, 1902, to Meany, 11 December 1902.
57. There is evidence that midwestern promoters were much more independent minded and aggressive than southern promoters. Paul Latzke, *A Fight with an Octopus* (Chicago, 1906), 53. See also AT&T Archives, box 1337, Interstate Independent Telephone Association, 1904. *Western Electrician,* 8 June 1907.

58. Even in New England, where there was little competition, the relatively rural states of Maine, Vermont, and New Hampshire achieved a ratio of 67.
59. See sources for table 5.3. The telephone figure is from the telephone census and the population figure is from the population census.
60. Fisher, "Revolution," p. 8.
61. Bureau of the Census, *Telephones and Telegraphs, 1902,* table 38.
62. Southern income per capita remained 51 percent of the national average between 1880 and 1900, while southern agricultural output per worker hovered around 56 percent. See Woodman, "Sequel to Slavery," 526, and Stanley Engerman, "Some Economic Factors in Southern Backwardness in the Nineteenth Century," in John Kain and J. Meyer, eds., *Essays in Regional Economics* (Cambridge, Mass., 1971), 303. See also Easterlin, "Interregional Differences in Per Capita Income, 1840–1950," tables A–1, A–2.
63. With measured service, customers could have paid for only the amount of service they wanted, thus allowing those who could not afford the flat rate to take service. In California, one of the first states to use measured rates, messages per capita were high, as in the Midwest, reflecting the wide distribution of telephones; but messages per telephone were much lower, reflecting the higher marginal cost of using the telephone. Bureau of the Census, *Telephones and Telegraphs, 1902,* tables 26, 31.
64. Compare Sidney Glazer, "The Rural Community in the Urban Age," *Agricultural History* 23, 2 (1949): 130–34. Also Howard Rabinowitz, "Continuity and Change: Southern Urban Development, 1860–1900," in David Goldfield and Blaine Brownell, eds., *The City in Southern History* (Port Washington, N.Y., 1977).
65. There are no comparable figures for 1902. See Fisher, "Revolution," for a discussion of the different ways in which different classes of customers used telephones.
66. AT&T Archives, box 1285, Telephone Service for Small Exchanges, French-Beach, 8 May 1894.
67. AT&T Archives, box 1011, Herbert Laws Webb, "Long Distance Telephony," *Electrical Engineer,* 4, 11 May 1892. At the start of competition, local and long-distance service had not been integrated. Perhaps the clearest evidence of this is the existence of a national long-distance directory for 1894 (AT&T Archives). Subscribers who used premium long-distance service rented a special high-powered transmitter in addition to a regular telephone, and used separate circuits. It was possible to make shorter calls over reg-

ular lines, but very long-distance communications required this special arrangement.

68. For a discussion of the reasons for and problems of creating a unified network, see AT&T Archives, box 1285, Toll Line Service, 1897–98, Doolittle-Hudson, 22 September 1898; 3 June 1898; 15 July 1898. AT&T Archives, box 1011, Building Early Long Distance Lines, Hall-Hudson, 21 January 1886.

Chapter 6. The Extension of the Network

1. Langdale, "The Growth of Long Distance Telephony."
2. For information on the economics of the telephone industry, see Robert Bornholz and David Evans, "The Early History of Competition in the Telephone Industry," in David Evans, ed., *Breaking Up Bell: Essays on Industrial Organization and Regulation* (New York, 1983), 7–40.
3. Bornholz and Evans, "The Early History," 12–15. McMeal, *The Story of Independent Telephony,* 81–84. As shown in chap. 5, the Interstate Telephone Company in North Carolina successfully fended off Bell incursions in this way. In the Midwest, associations of independent firms, as well as several larger companies, built similarly successful regional networks.
4. As we shall see, Bell made the first move to interconnect and compromise with independents in the South and similar regions. The notion that Bell's refusal to interconnect was a potent competitive weapon is an article of faith in telephone literature. See Brooks, *Telephone,* 114; Rhodes, *Beginnings of Telephony*; and Joseph Goulden, *Monopoly* (New York, 1968), 11.
5. E. J. Hall, "Long Distance Telephone Network," paper presented before the National Telephone Exchange Association Meeting, 26 September 1887, reprinted in the *Electrical Review* (1 October 1887). AT&T Archives, box 1011, Building Early Long Distance Lines, Hall-Hudson, 21 January 1886.
6. AT&T Archives, box 1375, Sub-license Policy, Leverett-Fish, 17 October 1901.
7. AT&T Archives, box 1330, Toll Line Service, 1901, Doolittle-Cochrane, 16 January 1901.
8. AT&T Archives, box 1285, Toll Line Service, 1892–96, Doolittle-Davis, 4 June 1896.

9. AT&T Archives, box 1057, T. B. Doolittle, Toll Lines, 1904 Bulletin to Operating Companies.
10. Ibid. See also AT&T Archives, box 1259, Long Distance Service, Extension, 1894, Doolittle-Hudson, 8 March 1894; 9 March 1894; 10 March 1894.
11. AT&T Archives, box 1285, Toll Line Service, 1897–98, Doolittle-Hudson, 22 September 1898; 3 June 1898; 15 July 1898.
12. AT&T Archives, box 1057, T. B. Doolittle, Toll Lines, 1904 Bulletin to Operating Companies.
13. AT&T Archives, box 1279, Long Distance and Toll Line Service in New England, 1894–95. AT&T Archives, box 1285, Toll Line Service, 1897–98, Doolittle-Hudson, 3 June 1898.
14. AT&T Archives, box 1285, Toll Line Service, 1897–98, Doolittle-Hudson, 19 May 1898.
15. AT&T Archives, box 1259, Long Distance Service, Extension, 1894, Doolittle-Hudson, 10 March 1894. AT&T Archives, box 1057, T. B. Doolittle, Toll Lines, 1904 Bulletin to Operating Companies.
16. Southern Bell reported a negative net income of $11,000 in 1901, $5,000 in 1902, and nearly $200,000 in 1904. In 1895, it had a positive net income of almost $125,000. SBT&T *Annual Reports*. SBT&T Corporate History, "Analysis of Surplus."
17. Garnet, *The Telephone Enterprise*, 120–21. AT&T Archives, box 1337, Central Union, Report on.
18. SBM, 1 January 1894; 2 October 1895.
19. SBM, 14 September 1892.
20. SBM, 20 February 1894.
21. SBM, 28 March 1895.
22. SBM, 14 November 1894.
23. AT&T Archives, box 1285, Telephone Service for Small Exchanges, SBT&T-French, 31 July 1894; McCluer-Carson, 13 April 1894. McCluer noted that every little town wanted service, generally at prices lower than Bell's.
24. AT&T Archives, box 1284, Party Line Development, 1898–99, Sykes-Durant, 11 March 1899.
25. Brock, *The Telecommunications Industry*, 117.
26. AT&T Archives, box 1285, Toll Line Service, 1892–96, Doolittle-Davis, 4 June 1896, 7 July 1896.
27. Ibid.
28. AT&T Archives, box 1340, SBT&T Headquarters Office Moved to Atlanta, Doolittle-Cochrane, 11 February 1901. AT&T Archives,

box 1285, Toll Line Service, 1897–98, Doolittle-Hudson, 1 November 1898; 23 November 1898.

29. AT&T Archives, box 1285, Toll Line Service, 1897–98, Doolittle-Hudson, 1 November 1898.
30. Ibid.
31. AT&T Archives, box 1285, Telephone Service for Small Exchanges, SBT&T-French, 31 July 1894, 7 October 1894. The cheap service which independents offered quickly undercut the price of Bell's inexpensive class C service.
32. AT&T Archives, box 1285, Toll Line Service, 1897–98, Doolittle-Hudson, 1 November 1898.
33. In the Central Union Company territory, Chicago Telephone and Telegraph ran Chicago, Cleveland Telephone and Telegraph took responsibility for Cleveland, and Central District and Printing Company held Pittsburgh.
34. AT&T Archives, box 1285, Toll Line Service, 1897–98, Doolittle-Hudson, 1 November 1898, 23 November 1898.
35. Ibid.
36. AT&T Archives, box 1330, Toll Line Construction, Doolittle-Hudson, 10 April 1899, 24 January 1899.
37. AT&T Archives, box 1285, Toll Line Service, 1897–98, Doolittle-Hudson, 1 November 1898.
38. Ibid.
39. AT&T Archives, box 1056, T. R. Lockwood, From the Beginning, 1926, p. 18. Lockwood regarded Hall as "[t]he most far seeing, all around competent and efficient telephone man of his day." Garnet, *The Telephone Enterprise,* 75.
40. SBT&T *Annual Report,* 1896.
41. SBM, 2 January 1896; 8 December 1897; 23 September 1896.
42. AT&T Archives, box 1039, Measured Service vs. Fixed Rates, 1880–98, Hall-Hudson, 10 December 1898. Hall advocated a new system of cost accounting to measure system costs.
43. Plans for this work were set forth in SBT&T *Annual Report,* 1896.
44. AT&T Archives, box 1340, Southern Bell Exchanges Report, 1900, Hall-Cochrane, 2 November 1900. See also AT&T Archives, box 1263, Southern Bell Financing, 1898–99, Hall-Hudson, 20 April 1898.
45. AT&T Archives, box 1340, Southern Bell Exchanges Report, 1900, Hall-Cochrane, 2 November 1900.
46. Ibid.
47. SBT&T Corporate History, "Stock Digest."

48. SBT&T Corporate History. Also AT&T Archives, box 1263, Southern Bell Financing, 1898–99, Hall-Hudson, 20 April 1898; 3 August 1898; 1 February 1899; 23 February 1899; 26 May 1899.
49. SBM, 1897–98.
50. SBT&T *Annual Report,* 1897.
51. SBT&T *Annual Report,* 1898.
52. SBM, 17 August 1898.
53. SBT&T *Annual Report,* 1899.
54. SBT&T *Annual Report,* 1904.
55. SBT&T *Annual Report,* 1908. Also, SBT&T Corporate History, "Stock Digest."
56. AT&T Archives, box 1340, Southern Bell Financing, 1901–10, Hall-Cochrane, 9 May 1901. AT&T Archives, box 1340, Southern Bell Exchanges Report, 1900, Hall-Cochrane, 2 November 1900.
57. AT&T Archives, box 1340, Southern Bell Financing, 1901–10, Hall-Fish, 14 November 1902. AT&T Archives, box 1340, Southern Bell Stock Purchased from Western Union, 1902, Hall-Fish, 24 July 1902.
58. AT&T Archives, box 1340, Southern Bell Financing, 1901–10, Hall-Vail, 14 February 1908; 14 July 1910.
59. SBT&T Corporate History, "Capital Stock Issued and Dividends Paid."
60. AT&T Archives, box 1340, SBT&T Acquisition of Independent Companies, 1897–1901, Hall-Fish, 8 October 1901.
61. AT&T Archives, box 1340, SBT&T Acquisition of Exchanges in North Carolina, 1903, Hall-Fish, 24 October 1903.
62. GMLB 523, pp. 272–92.
63. AT&T Archives, box 1340, Southern Bell Exchanges Report, 1900, Hall-Cochrane, 2 November 1900.
64. Ibid.
65. SBM, 26 January 1895.
66. AT&T Archives, box 1340, SBT&T Acquisition of Independent Companies, 1897–1901, Hall-Hudson, 21 February 1898.
67. Ibid., 10 November 1899; Williams-Merrihew, 18 February 1898.
68. AT&T Archives, box 1033, Richmond, Virginia, "The Telephone In Virginia," *Richmond News Leader* (31 December 1925).
69. AT&T Archives, box 1340, SBT&T Acquisition of Independent Companies, 1897–1901, Easterlin-Carson, 17 May 1897.
70. Ibid., 3 June 1897; also Barnwell-Easterlin, 1 June 1897; Hall-French, 16 June 1897.
71. AT&T Archives, box 1340, Georgia Independents Acquired by

SBT&T, 1902–11, Gentry-French, 30 October 1902.

72. AT&T Archives, box 1340, SBT&T Acquisition of Independent Companies, 1897–1901, Hall-Cochrane, 6 March 1901.
73. Ibid. See also Hall-Hudson, 19 May 1897.
74. Ibid.
75. Hall wrote that the plant of both Southern Bell and its competitor were in "wretched" condition. Nonetheless, he believed that the city would be a valuable toll line center. Ibid., 5 October 1897.
76. Ibid.; Report of J. M. Brown, 20 October 1897. Instances of head-to-head competition in the same town were not common.
77. Ibid., Hall-Fish, 14 December 1901.
78. SBM, 4 June 1903; 9 November 1903; 8 November 1906; 9 May 1907; 4 April 1907; 10 January 1907.
79. AT&T Archives, box 1263, Sub-License Contracts, 1898–99, Wilson-Hall, 9 February 1899.
80. Contrast Brock, *The Telecommunications Industry,* 110, 155–57. AT&T Archives, Illinois State Regulation, Sunny-Kingsbury, 6 July 1915. Independents often resisted incorporation into the Bell network for fear of being subsumed by the larger company. See AT&T Archives, box 1337, Independent Telephone Company Association, 1901–2. See also Gabel, "The Early Competitive Era in Telephone Communications," 350, 353–54.
81. SBM, 22 October 1902; 9 November 1903; 23 February 1906.
82. AT&T Archives, box 1263, Sub-License Contracts, 1898–99, Easterlin-Wilson, 26 September 1898.
83. AT&T Archives, box 1340, North Carolina Independents Acquired by SBT&T, 1902–11, Agreement with Asheville Telephone Company, 16 July 1903.
84. AT&T Archives, box 1263, Interconnection with Southern States Telephone and Telegraph, Hall-Hudson, 8 October 1897.
85. Ibid., Hall-French, 30 August 1897; Memorandum, 16 April 1897.
86. AT&T Archives, box 1340, North Carolina Independents Acquired by SBT&T, 1902–11, Agreement with Asheville Telephone Company, 16 July 1903. See also AT&T Archives, box 1263, Acquisition of Citizens Telephone Company of Pensacola, Florida, 1895. By forcing newly subordinated firms to adhere to Bell practices, Bell made it easier to extend its system to remote places formerly under independent control.
87. AT&T Archives, box 66, Sub-Licensing Policy, 1907–8, Hall-French, 16 May 1908.
88. AT&T Archives, box 1263, Interconnection with Southern States

Telephone and Telegraph, Hall-Hudson, 8 October 1897.

89. AT&T Archives, box 1285, Telephone Service for Small Exchanges, 1894, Beach-French, 12 April 1894; French-Beach, 8 May 1894.
90. AT&T Archives, box 66, Sub-Licensing Policy, 1907–8, Hall-French, 16 May 1908.
91. Ibid.
92. AT&T Archives, box 1348, Sub-Licensing, Advantages to Operating Companies, 1903, Hall-Fish, 31 July 1903.
93. Ibid.
94. Ibid., Davis-Fish, 11 August 1903.
95. AT&T Archives, box 1348, Operating Company Investment in Branch Line Company, 1904, Wallace-Fish, 11 October 1904; Fish-Wallace, 17 October 1904.
96. Gabel, "The Early Competitive Era," 351.
97. AT&T Archives, box 1364, Sub-License Statistics, Vail Circular Letter, 9 October 1907.
98. AT&T Archives, box 1263, Interconnection with Southern States Telephone and Telegraph, Easterlin-Carson, 15 April 1897.
99. AT&T Archives, box 1340, North Carolina Independents, Agreement with Asheville Telephone Company, 16 July 1903. See also AT&T Archives, box 1263, Sub-License Contracts, 1898–99, Hall-Hudson, 10 February 1899; Wilson-Hall, 9 February 1899; Hall-Hudson, 14 February 1899.
100. AT&T Archives, box 1263, Sub-License Contracts, 1898–99, Hall-Hudson, 10 February 1899; Wilson-Hall, 9 February 1899.
101. AT&T Archives, box 1364, Sub-License Statistics, Vail Circular Letter, 10 February 1908. The liberalization of sublicensing did not mean that AT&T was willing to allow competing long-distance companies into its network. Here company policy continued to prohibit interconnection. See Gabel, "The Early Competitive Era," 350–53.
102. AT&T Archives, box 1263, Sub-License Contracts, 1898–99, Hall-Hudson, 10 February 1899; Wilson-Hall, 9 February 1899.
103. AT&T Archives, box 1340, North Carolina Independents Acquired by SBT&T, 1902–11, Agreement with Asheville Telephone Company, 16 July 1903.
104. AT&T Archives, box 1277, Exchange Contract, Sub-License, 1899.
105. Garnet, *The Telephone Enterprise,* 128–30.
106. AT&T Archives, box 1010, Bell System, Possible Organization

Plans, Hall-Bowditch, 22 February 1886. Garnet, *The Telephone Enterprise,* 83–85.

107. AT&T Archives, box 1057, T. B. Doolittle, Toll Lines, 1904 Bulletin to Operating Companies. For Hall's ideas on central coordination, see AT&T Archives, box 1010, Bell System, Possible Organization Plans, Hall-Hudson, 7 January 1889. This letter and the one to Bowditch of 22 February 1886 show Hall wavering between central control and decentralized operations. He continued to move between these two poles at Southern Bell, finally striking what I believe was the proper balance between the two.
108. SBT&T *Annual Reports,* 1883, 1901.
109. AT&T Archives, box 1377, AT&T–Operating Company Relations, Hall-Fish, 5 July 1906.
110. Ibid.
111. Ibid.
112. Ibid. Also AT&T Archives, box 1309, Industrial Commission Hearings Report, E. J. Hall, 1901.
113. AT&T Archives, box 1377, AT&T–Operating Company Relations, Hall-Fish, 5 July 1906.
114. Ibid.
115. Alfred D. Chandler, Jr., *Strategy and Structure: Chapters in the History of American Industrial Enterprise* (Cambridge, Mass., 1962), 163–70. Chandler discusses the problems of centralized structure at Standard Oil. Judging from Hall's later views on functional organization, it seems that he was aware of the problem. See also, Garnet, *The Telephone Enterprise,* 139–40.
116. AT&T Archives, box 1377, AT&T–Operating Company Relations, Hall-Fish, 5 July 1906.
117. This basic rule for long-distance development Hall had stated earlier. The economic logic of system building required that Bell apply its network first in the most profitable centers of demand and then gradually expand outward and fill in the gaps. Competitive pressures had apparently erased the memory of many Bell managers. See AT&T Archives, box 1011, Building Early Long Distance Lines, Hall-Hudson, 21 January 1886.
118. Bureau of the Census, *Telephones and Telegraphs, 1912,* table 28, p. 38.
119. Brooks, *Telephone,* 102–7.
120. Noobar T. Danielian, *AT&T: The Story of Industrial Conquest* (New York, 1939), 46–60. Kolko, *The Triumph of Conservatism,* 47–49.

121. The two exceptions to this pattern, New England and the Midwest, actually confirm the rule. In Bell-dominated New England, telephone distribution had been too low, thus it jumped 3.4 times between 1902 and 1912. In the Midwest, distribution only increased 2.7 times in the same period. But telephone distribution in the region had grown markedly after 1894, with the advent of competition. As Bell reestablished market control, midwestern growth slowed down, dipping below the national average.
122. Conflict between large-scale organizations and local interests is a common theme in late-nineteenth-century American history. See Larson, *Bonds of Enterprise,* chap. 5.

Chapter 7. A Merging of Interests

1. Bureau of the Census, *Telephones, 1912,* table 11, p. 24. Wire mileage is an imperfect measure of expansion. The West and Southwest naturally experienced great growth in wire mileage, as distances between exchanges were much greater there than elsewhere. The census definition of the Southeast excludes Alabama. Nationally Bell and independent wire mileage increased by the same percentage. See also United States Bureau of the Census, *Telephones and Telegraphs, 1907* (Washington, D.C., 1909), table 8.
2. SBT&T Sect., Pickernell—Summary of Hoxsey Report to Hall, 21 March 1908. AT&T Archives, box 2025, Toll Line Statistics, 1906.
3. AT&T Archives, box 2026, Southern Bell Telephone and Telegraph Company Toll Traffic Matters, Cotton-Carty, 14 July 1909. See also AT&T Archives, box 1285, Telephone Service for Small Exchanges, Gentry-French, 31 July 1894.
4. SBT&T Sect., Pickernell-Hall, 13 March 1908.
5. Fisher, "The Revolution in Rural Telephony," 8, claims that by 1912 the percentage of farmers with telephones matched that for urbanites; by 1920 farmers had surpassed urban residents. The percentage of southern farms with telephones was much lower, however.
6. AT&T Archives, box 1372, Rural Telephone Service, 1905–6, Pickernell-Fish, 20 February 1906.
7. GMLB 605, Smith-Bell Company, 3 May 1901. AT&T Archives, box 1342, Rural Telephone Service, 1903–4. AT&T Archives, box 1214, Rural Telephone Service, 1899–1902, Central Union General Order #7.

8. AT&T Archives, box 1285, Telephone Service for Small Exchanges, Gentry-French, 31 July 1894.
9. AT&T Archives, W. S. Allen Papers, Allen-Fish, 16 May 1906.
10. AT&T Archives, President's Letterbook (hereafter PLB) 27, Fish-Trowbridge, 8 April 1903. PLB 28, Fish-Gentry, 21 April 1905.
11. PLB 39, Fish-Jackson, 9 May 1905.
12. PLB 34, Fish-Bruehler, 30 April 1904.
13. AT&T Archives, box 1372, Rural Telephone Service, 1905–6, Pickernell-Fish, 20 February 1906.
14. As early as 1901, Western Electric personnel had advised Fish to move in this direction, but he would not listen. AT&T Archives, box 1332, Western Electric Cheaper Equipment for Smaller Exchanges, Barton-Fish, 21 November 1901.
15. AT&T Archives, box 1372, Rural Telephone Service, 1905–6, Hayes Circular Letter to All Operating Companies, 10 May 1906.
16. Ibid.
17. Ibid.
18. Ibid. AT&T Archives, box 1329, Switchboard Development, 1905–7, Paxson-Hayes, 25 August 1906.
19. The achievement was short-lived. When competitive pressure subsided, Bell again lost interest in rural telephony. See Claude Fisher, "Technology's Retreat: The Decline of Rural Telephony in the United States, 1920–1940," *Social Science History* 11, 3 (1987): 295–327.
20. Quoted in Georgia Railroad Commission, *Annual Report,* 1909, pp. 32–35.
21. AT&T Archives, box 1363, Southern Bell Rural Telephone Service, 1907–10, Harris-Ellsworth, 30 June 1909.
22. Ibid., Shepard–General Contract Agent, 14 April 1908.
23. Quoted in Georgia Railroad Commission, *Annual Report,* 1909, pp. 32–35.
24. Ibid. The increase took place between December 1907 and March 1909, as the number of rural stations rose from 1,961 to 5,731. Between 1902 and 1907, rural stations in the South Atlantic (all Southern Bell states except Alabama, but including Maryland and Washington, D.C.) connected to commercial lines increased by 5.7 times, a factor greater than that of the national average of 5.5. See Bureau of the Census, *Telephone Census, 1907,* table 12. See also SBT&T Sect., Pickernell-Hall, 13 March 1908.
25. Bureau of the Census, *Telephone Census, 1912,* table 11. Almost 75 percent of wire mileage was Bell's by 1912. Clearly the decline

of competition and reduction in duplicate facilities had much to do with the slower expansion, but these same factors operated nationally, so they cannot explain why the South fell far behind relative to other regions.

26. SBT&T Sect., Pickernell—Summary of Hoxsey Report to Hall, 21 March 1908. See also Hall-Gentry, 13 March 1908; 23 September 1908.
27. SBT&T Sect., Pickernell—Summany of Hoxsey Report to Hall, 21 March 1908.
28. Ibid.
29. Ibid.
30. In telephone cables with two circuits (pairs of wires) it was possible to create a third circuit without the addition of any wire. When the two circuits simultaneously carried separate telephonic messages, a third message could be transmitted superimposed on the two. This third message was said to be carried by the "phantom" circuit. The obvious advantage of this arrangement is that it increased capacity without the expense of adding more wire circuits to cables. See Rhodes, *Beginnings of Telephony,* 189–91.
31. SBT&T Sect., Hall-Gentry, 23 September 1908.
32. AT&T Archives, box 2026, Southern Bell Telephone and Telegraph Company, Toll Traffic Matters, Cotton-Carty, 14 July 1909. Cotton noted a year after Hoxsey's report that more capacity for toll lines was still needed and rates had to be raised on local service.
33. Ibid., Hall-Vail, 22 September 1908.
34. This issue is discussed in depth in chap. 8.
35. AT&T Archives, box 2026, Southern Bell Telephone and Telegraph Company, Toll Traffic Matters, Cotton-Carty, 14 July 1909, pp. 5–6.
36. Ibid., pp. 4–5.
37. Ibid., pp. 12–13.
38. Ibid., p. 5.
39. Ibid., pp. 9–11.
40. Ibid., p. 11.
41. AT&T Archives, box 1364, Sub-Licensee Statistics. Overall only 7.7 percent of the sublicensees were owned by operating companies.
42. AT&T Archives, box 2026, Southern Bell Telephone and Telegraph Toll Traffic Matters, Cotton-Carty, 14 July 1909.
43. AT&T Archives, box 1372, Rural Telephone Service, 1905–6, Hayes Circular Letter to All Operating Companies, 10 May 1906,

p. 10. Hayes meant that more labor, proportionally, was consumed in rural operations than in others; or in other words, the average productivity of labor was lower.

44. Garnet, *The Telephone Enterprise,* 112, 128.
45. AT&T Archives, box 1377, Committee to Reconsider Organization of AT&T, 1907. PLB 51, Vail–Crane et al., 19 March 1908.
46. Garnet, *The Telephone Enterprise,* 128.
47. Ibid., 136–37.
48. AT&T Archives, box 2029, Development of Functional Organization in the Bell System, 1895–1935, H. F. Thurber, "The Necessity of Cooperation in Our Work," 8 March 1904; see also "An Application of Some General Principles of Organization," 1909.
49. Garnet, *The Telephone Enterprise,* 137–38.
50. PLB 51, Vail–Crane et al., 19 March 1908.
51. AT&T Archives, Fish Private Letterbook, 48, Vail-Hall, n.d.
52. PLB 51, Vail-Hall, 14 February 1908.
53. SBT&T *Annual Report,* 1909.
54. AT&T Archives, box 2029, Development of Functional Organization in the Bell System, 1895–1935, Thayer-Vail, 25 January 1910.
55. Ibid., Central Union Telephone–Carty, 10 June 1908; Pacific Telephone–Carty, 27 June 1908.
56. Ibid., New York Telephone–Carty, 1 March 1899.
57. AT&T Archives, box 1017, New England Telephone and Telegraph Company, Organization of Departments and Duties, Organizational Charts, 1892, 1898, 1902, 1906, 1908.
58. AT&T Archives, box 2029, Development of Functional Organization in the Bell System, 1895–1935, Houston-Carty, 25 May 1908.
59. Ibid., Watson-Carty, 5 January 1910.
60. Ibid., Fields-Carty, 8 September 1908.
61. Ibid., Hall-Vail, 27 September 1909, p. 5.
62. AT&T Archives, box 1010, Bell System, Possible Organization Plans, Hall-Bowditch, 22 February 1886.
63. Ibid., Hall-Hudson, 7 January 1889.
64. Garnet, *The Telephone Enterprise,* 137–39.
65. AT&T Archives, box 2029, Development of Functional Organization in the Bell System, 1895–1935, Fields-Carty, 8 September 1908.
66. Ibid., Hall-Vail, 27 September 1909. For his earlier thoughts on these matters, see AT&T Archives, box 1011, Building Early Long Distance Lines, Hall-Hudson, 21 January 1888.
67. AT&T Archives, box 2029, Development of Functional Organiza-

tion in the Bell System, 1895–1935, Hall-Vail, 27 September 1909, p. 1.

68. Ibid., p. 2.
69. Ibid., p. 4.
70. AT&T Archives, box 1035, Engineering Department Organization and Function, 1894–99, Davis-Hudson, 28 March 1899. AT&T obtained greater cooperation from operating companies by sending its own agents to them with advice on technology.
71. AT&T Archives, box 1315, Standardization of Apparatus, 1899, Davis-Hudson, 18 April 1899. AT&T Archives, box 1305, Technical Department, Increase in Personnel, 1899–1900, Hayes-Hudson, 8 March 1900; 28 September 1899. AT&T engineers rather than engineers at individual operating companies supplied models and designs. AT&T representatives also moved to the field to assist operating companies in setting up long-distance lines, designing facilities for system expansion, and inspecting equipment to make sure it conformed to system specification. Garnet, *The Telephone Enterprise,* 120–24. See also AT&T Archives, box 1341, Engineering Department, Organization and Duties, 1901–3. Local engineers reported to AT&T personnel.
72. Garnet, *The Telephone Enterprise,* 123–24. Only about half of the companies participated in the effort to link local engineers to AT&T technical departments.
73. Ibid.
74. AT&T Archives, box 1341, Engineering Department Organization and Duties, 1901–3, Boyce-Davis, 26 July 1902; Davis-Fish, 29 March 1902; Response to Form Circular, 2 April 1902.
75. Ibid., Davis-Fish, 11 February 1903. For a discussion of this problem earlier, see AT&T Archives, box 1282, Equipment, Quality Control, Western Electric, 1895, J. N. Keller–French, 1 April 1895; Farnham-Keller, 7 March 1895; see also Hayes-French, 4 April 1895.
76. AT&T Archives, box 1341, Engineering Department, Organization and Duties, 1901–3, Davis-Fish, 11 February 1903. Also AT&T Miscellaneous French Letterbooks, 1903, French-Fish, 10 February 1903. AT&T Archives, box 1361, Apparatus Development Committee, 1904, Davis-Fish, 31 May 1904; Committee Report, 10 April 1903.
77. AT&T Archives, box 1341, Engineering Department Organization and Duties, 1901–3, Davis-Fish, 11 February 1903.
78. Garnet, *The Telephone Enterprise,* 123.

79. AT&T Archives, box 1333, Exchange Development Plans, 1902, Davis-Fish, 28 August 1902. Even small city exchanges were advised to plan ahead, as continued reductions in competition would eliminate fluctuations in demand and allow them to upgrade equipment to system standards. AT&T Archives, box 1011, Development Plans, Small Cities, 1905.
80. AT&T Archives, box 1341, Engineering Department Organization and Duties, 1901–3, Davis-Fish, 11 February 1903.
81. AT&T Archives, box 1361, Apparatus Development Committee, 1904, Davis-Fish, 10 April 1904; 31 May 1904; Barton-Fish, 2 June 1904.
82. Garnet, *The Telephone Enterprise,* 137.
83. With responsibility for planning and design centralized in New York and AT&T restructured along functional lines, Bell overcame a major impediment to standardization. AT&T Archives, box 1377, AT&T Company Headquarters Department, Hayes-Vail, 24 May 1907.
84. AT&T Archives, box 1377, AT&T Company Headquarters Department, Carty-Hall, 17 July 1907; Hall-Carty, 19 July 1907; 20 March 1908.
85. AT&T Archives, box 1008, ABT Ownership of Operating Company Stock, 1880–90. AT&T Archives, box 1302, ABT Ownership of Operating Company Stock, 1897. AT&T Archives, box 2054, Purchase of Operating Company Stocks and Bonds, 1880–1904. Garnet, *The Telephone Enterprise,* 106–7, 116–17, 133.
86. Garnet, *The Telephone Enterprise,* 113–16. AT&T Archives, box 1377, Operating Companies and AT&T Financial Report, 1900–1905.
87. Garnet, *The Telephone Enterprise,* 132–33.
88. AT&T Archives, box 66, Sub-Licensing Policy, 1907–8, Vail Circular Letter, 10 February 1908; Vail-Watson, 9 October 1907. This policy, the president wrote, would allow Bell firms to reduce outlays. See also Vail-Yost, 4 February 1908; 5 February 1908.
89. Ibid., Hall-French, 16 May 1908.
90. Between 1907 and 1908 the percentage of independent telephones connected to the Bell system leaped from 13.7 to 26.6; United States Bureau of the Census, *Historical Statistics from Colonial Times to 1970* (Washington, D.C., 1975), 2:783.
91. Garnet, *The Telephone Enterprise,* 152–54.
92. AT&T Archives, box 15, Agreement with Independent Companies on Toll Service, Kingsbury-Guernsey, 29 December 1914.

93. AT&T Archives, box 1364, Sub-Licensee Statistics, Vail-Watson, 9 October 1907.
94. AT&T Archives, box 66, Sub-Licensing Policy, 1907–8, Vail-Watson, 10 September 1908. See also Hall-French, 16 May 1908; Vail Circular Letter, 10 February 1908.
95. AT&T Archives, box 8, Long Lines, Operation by Associated Companies, 1912, Rorty-Thayer, 21 February 1912; Gifford-Thayer, 9 February 1912. AT&T Archives, box 4, Organizational Matters, Long Lines, 1912–20, Stevenson-Kingsbury, 10 June 1912.
96. AT&T Archives, box 2029, Development of Functional Organization in the Bell System, 1895–1935, "An Application of Some General Principles of Organization," 1909.
97. AT&T Archives, box 2026, Southern Bell Telephone and Telegraph Company Toll Traffic Matters, Cotton-Carty, 14 July 1909.
98. AT&T Archives, box 21, Southern Bell Telephone and Telegraph Company, Acquisition of Independent Firms, Policy, 1912, Remarks of W. T. Gentry at New York Conference, 27 March 1912. Southern Bell renewed efforts at acquisition. The Kingsbury Commitment only banned acquisition of competing companies. Even these companies could be purchased, moreover, with permission from the government. Bell thus continued its policy of buying independent companies. Garnet, *The Telephone Enterprise*, 152–54.

Chapter 8. A Role for Regulation

1. AT&T Archives, box 1265, ABT Operation, 1880–84, Vail-Forbes, 25 January 1883. Vail discusses the early legislation.
2. Brock, *The Telecommunications Industry*, chap. 6. Brock takes a different view, seeing competition as the main issue of regulation.
3. SBM, 3 July 1895.
4. AT&T Archives, Columbus, Georgia, Contract with Southern Bell, 6 September 1900; Newman, Georgia, Ordinance, 17 December 1906. GMLB 631, Sharp-French, 26 August 1901.
5. *Montgomery Advertiser*, 11 November 1898.
6. *Huntsville Mercury*, 9 December 1899. *Birmingham Age Herald*, 7 February 1899.
7. AT&T Archives, Lake City, Florida, Ordinance, 10 May 1906.
8. AT&T Archives, Interstate Commerce Commission Investigation, SBT&T, Documents Prepared for N. C. Kingsbury, Tampa Franchise Agreement, 7 June 1899.

9. AT&T Archives, W. T. Gentry Special Report, Florida, 7 March 1910.
10. AT&T Archives, SBT&T, Florida, Gentry-Adams, 11 December 1901. Apex, North Carolina, Agreement with Southern Bell, 1 May 1913. Southport, North Carolina, Agreement with Southern Bell, 1909. Cary, North Carolina, Petition to North Carolina Corporation Commission, 11 February 1915. Salisbury, North Carolina, Petition to North Carolina Corporation Commission, 20 July 1910. Denton, North Carolina, Petition to North Carolina Corporation Commission, 12 November 1910. Greensboro, North Carolina, Resolution, 11 January 1904.
11. *Birmingham Age Herald,* 14 February 1899.
12. AT&T Archives, Valdosta, Georgia, Petition to Southern Bell Telephone and Telegraph Company, 15 April 1913.
13. Georgia Railroad Commission Report, 1910–11, vol. 2, Municipalities and Counties, pp. 25–27, Gainsboro Telephone Case at Carrollton.
14. AT&T Archives, Gibson, North Carolina, City Ordinance, 31 May 1910.
15. The number and diversity of interests in large cities could lead to internecine conflicts. Independent telephone promoters fought with powerful business interests, who demanded higher quality and more extensive service and were willing to ally with Bell to get it.
16. AT&T Archives, box 1148, Atlanta, Georgia Franchise, Meany-Fish, 6 September 1902. GMLB 923, 1907, pp. 78–112.
17. Ibid.
18. GMLB 923, 1907, pp. 78–112.
19. Ibid.
20. Ibid. Also *Atlanta Journal,* 7 April 1907.
21. SBM, 1 January 1888.
22. AT&T Archives, box 32, Southern Bell Acquisition of Atlanta Telephone and Telegraph, 1918–19.
23. Brock, *The Telecommunications Industry,* 112–13.
24. AT&T Archives, Richmond, Virginia, Ordinance, 26 June 1884.
25. The city was no doubt taking advantage of its new leverage, with the reintroduction of competition. AT&T Archives, box 1340, Acquisition of Independent Companies, 1897–1901, Hall-Hudson, 21 February 1898; 10 November 1899; 29 May 1899.
26. Southern Bell Telephone and Telegraph Company v. City of Richmond, Virginia, 4th Circuit Court of Appeals, 9 July 1900. See Federal Reporter, vol. 103, p. 31 (hereafter 103 Fed 31.)

27. AT&T Archives, Spencer, North Carolina, Ordinance, 8 January 1904; Wendell, North Carolina, Ordinance, 1 May 1912.
28. Gavin Wright, "Rethinking the Postbellum Southern Political Economy: A Review Essay," *Business History Review* 58, 3 (1984): 412–16. Larson, *Bonds of Enterprise*, 171–96. Wright notes the ambivalence of southern towns and cities regarding outside capital; Larson discusses the efforts of western towns to regulate railroads to suit their needs.
29. Fearing Bell competition, they wanted city officials to force Bell either to reduce its rates to their levels or to leave the city. AT&T Archives, W. T. Gentry Special Report, Florida, 7 March 1910.
30. Ibid.
31. AT&T Archives, SBT&T, Gentry-Wrens, 22 December 1904.
32. AT&T Archives, box 1331, Right of Way, City of Richmond, Hall-Hudson, 29 May 1899; 12 July 1899; 4 September 1899; Carter–Holladay & Fearons, 15 December 1899. AT&T Archives, box 1033, Richmond, Virginia, "The Telephone in Virginia."
33. In such places Bell had to concede a large hand to independent firms anyway, negating potential conflicts between representatives of local interests and the national firm.
34. AT&T Archives, box 1340, North Carolina Independents Acquired by SBT&T, 1902–11, Agreement with Asheville Telephone and Telegraph, 16 July 1903.
35. SBM, 10 January 1907.
36. AT&T Archives, Minutes of Capital City (North Carolina) Telephone Company, 28 December 1906; Raleigh, North Carolina, City Ordinance, 20 December 1906. The agreement was much like a regular Bell sublicense contract. The city also received five free telephones for its compliance.
37. AT&T Archives, box 21, Acquisition of Independent Telephone Companies, SBT&T Policy, 1912, W. T. Gentry Remarks, New York City, 27 March 1912.
38. AT&T Archives, Tampa, Florida, Ordinance, 1904; Gentry–Peninsular Telephone Company (Florida), 5 October 1905.
39. North Carolina State Historical Library, Raleigh, North Carolina, Biennial Message of Governor William Kitchen to the Georgia State Assembly, 12 January 1909.
40. AT&T Archives, Chicago City Council Proceedings, 9 November 1878; 4 October 1887.
41. AT&T Archives, "An Ordinance Granting Privileges to the Chicago Telephone Company, and An Ordinance Regulating Telephone

Charges in the City of Chicago," 6 November 1907.

42. AT&T Archives, "An Ordinance Fixing the Maximum Rates for Telephone Service, and An Ordinance Creating a Telephone Bureau," 14 June 1913.
43. AT&T Archives, box 13, Central Union Proposed Reorganization, 1910–13, Sunny-Kingsbury, 4 December 1912.
44. AT&T Archives, Chicago City Ordinances, 6 November 1907; 14 June 1913.
45. AT&T Archives, box 13, Central Union Proposed Reorganization, 1910–13, Sunny-Kingsbury, 4 December 1912.
46. AT&T Archives, box 1321, Right of Way, Telephone Companies, 1900–1901, Wilson-French, 28 February 1900.
47. Ibid., Richardson-Yost, 5 December 1900; Cook-Richardson, 3 December 1900; Richardson-Cochrane, 13 December 1900.
48. J. Warren Stehman, *The Financial History of the American Telephone and Telegraph Company* (Boston, 1925), 261–62. Stehman claims that state regulation was not important, but he misses the crucial conflict between Bell and municipal governments, which state regulators helped to end.
49. Douglas D. Anderson, *Regulatory Politics and Electric Utilities* (Boston, 1981), 33–56.
50. Gabel, "The Early Competitive Era," 358, concurs with this, though he seems to feel that Bell did have some direct influence over policy.
51. Brock, *The Telecommunications Industry,* chap. 6. William K. Jones, "Origins of the Certificate of Public Convenience and Necessity: Developments in the States, 1870–1920," *Columbia Law Review* 79, 3 (1979): 453, 478.
52. SBT&T *Annual Report,* 1902.
53. Modern rate regulation employs the concept of a rate base, or investment on which the regulated firm is entitled a specific return. The choice of rate base is a political decision. In telephony, with no clear way of separating local from system costs, the rate base cannot be determined on purely economic criteria.
54. Brock, *The Telecommunications Industry,* 158.
55. Ford P. Hall, "Certificates of Public Convenience and Necessity," *Michigan Law Review* 28, 2 (1929): 122–23. Jones, "Origins of the Certificate," 454–55.
56. Jones, "Origins of the Certificate," 429–30. Later, independent commissions would address these issues as well; but by then they had gained power over modern regulatory matters such as market structure, rates, and interconnection.

57. The original act was somewhat ambiguous, stating that the Railroad Commission could approve all municipal contracts with utilities for "fairness," and to be sure that they provided "good service." It also had some control over the management and financing of utilities. But the act specifically stated that the commission's powers did not supersede the right of municipalities to contract with service suppliers. AT&T Archives, box 1148, Georgia Railroad Commission Act, 1907.
58. North Carolina Corporation Commission, *Annual Report,* 1897, 1899. Even as late as 1911, the Corporation Commission upheld the right of cities to use franchises as a regulatory tool. See North Carolina Corporation Commission, *Annual Report,* 1911, p. 262.
59. Laws of Florida, 1911, chap. 6186, #67.
60. Florida Railroad Commission, Report, vol. 16 (hereafter 16 FLRRCR), p. 50, order #358, 17 May 1912. This order denied that the commission had power over municipal rates. Laws of Florida, 1913, chap. 6525, #105, sec. 28, seemed to confirm commission power over municipalities. See 16 FLRRCR 20.
61. Laws of Florida, 1913, chap. 6525, #105, secs. 8–10. In 1914 it ordered all telephone companies to cease municipal rebates and special low rates to city governments, even if they were written into the franchise agreement. 18 FLRRCR 23.
62. Laws of Florida, 1913, chap. 6525, #105, secs. 3–4, 15.
63. Ibid.
64. 20 FLRRCR 42, 45. The Daytona firm had to sell out to Bell.
65. 19 FLRRCR 23.
66. AT&T Archives, SBT&T, Florida, Ames-Laird, 7 March 1917; Barnet-Carty, 16 June 1911; Gentry–Peninsular Telephone Company (Florida), 22 February 1909.
67. Laws of Florida, 1913, chap. 6525, #105, secs. 12, 13.
68. Laws of Florida, 1913, chap. 6525, #105, sec. 16. The Florida law on telephone interconnections conformed to Bell policy. It allowed interconnections only when firms were not direct competitors; direct competitors would have benefited more than did Bell from interconnection.
69. Virginia, Acts of the Assembly, 1904, chap. 8, secs. 1294(H) and 1294(B).
70. Ibid., 1906, chap. 310, sec. 2.
71. Ibid., 1914, chap. 340.
72. AT&T Archives, Georgia State Acts, "An Act Increasing the Membership and Powers of the Railroad Commission," 22 October 1907.

73. Georgia Railroad Commission, *Annual Report,* 1910, p. 31.
74. AT&T Archives, Dawson, Georgia, Right of Way Franchise, 6 October 1908. The court on 28 October 1911 decreed that the Railroad Commission was responsible for deciding disputes between telephone firms and municipal governments. See *Southeastern Reporter,* vol. 72 (hereafter 72 SE), p. 508. Also Georgia Railroad Commission, *Annual Report,* 1911, p. 398; 1914, p. 242.
75. AT&T Archives, Georgia, Chipley-Gentry, 22 January 1914. Georgia Railroad Commission Telephone Cases, vol. 2, p. 11, general order #10, 8 January 1908.
76. True, Bell as the larger firm could have tried to use predatory pricing to defeat rivals. But this would have actually hurt Bell more than the other firms, for Bell had a system to support, while they did not. AT&T Archives, SBT&T, Florida, Burr-Chipley, 9 October 1914. Bell feared that rate wars would adversely affect the rate of return on its entire system.
77. Georgia Railroad Commission, *Annual Report,* 1914, p. 243. Georgia Railroad Commission Report, 1910–11, vol. 2, Municipalities and Counties, pp. 25–27.
78. AT&T Archives, A. T. Jones Complaint to the Georgia Railroad Commission, 6 August 1911.
79. Georgia Railroad Commission Report, 1910–11, vol. 2, Municipalities and Counties, p. 64.
80. AT&T Archives, SBT&T, Georgia, Chipley-Gentry, 19 November 1906.
81. Significantly, the law which gave the Georgia Railroad Commission power over telephones stopped short of interfering with the internal affairs of Bell. As a foreign corporation, SBT&T was not subject to the commission's powers of review of finances. Local, non-Bell firms, of course, were. See Georgia Railroad Commission, Orders, Rulings and Decisions Directly Affecting Telephone and Telegraph Companies, part 1, 4 June 1912.
82. Address of C. Murphey Candler, reprinted in the *Southern Telephone News,* 9 March 1915.
83. Georgia Railroad Commission, *Annual Report,* 1914, p.14.
84. AT&T Archives, box 2019, SBT&T Progress Report, 1911, Gentry-Bethell, 13 April 1911.
85. The states with weak commissions generally had laws that limited their powers to making rules for the "safety and convenience" of patrons. These powers were insufficient to overcome municipal regulation.
86. North Carolina Acts, 1899, chaps. 512, 513. North Carolina Acts,

1891, chap. 391. North Carolina Acts, 1899, chap. 164.

87. North Carolina Acts, 1899, chap. 164, sec. 22, pt. 10. This section prevented prohibition of free municipal service.
88. North Carolina Acts, 1899, chap. 164, secs. 2, 12.
89. North Carolina Acts, 1899, chap. 164, sec. 2, pt. 23; sec. 21. Sections 23 and 13 of the law may have redressed this deficiency.
90. Mayo v. Western Union, North Carolina Supreme Court, 7 March 1897, 16 SE 1006. See also sections 18, 24 of the 1899 law, which further weakened the commission's control over telephones.
91. North Carolina Acts, 1907, chap. 74. See also chaps. 469, 966.
92. Ibid.
93. Clinton-Dunn v. Carolina Telephone and Telegraph, North Carolina Supreme Court, 17 April 1912, 74 SE 636. The court had to push the commission to take more responsibility over telephones. The commission in 1911 refused to hear a complaint by citizens of Asheboro against Southern Bell, noting that the town should use its franchise power to obtain relief. North Carolina Corporation Commission, *Annual Report*, 1911, pp. 262–63.
94. North Carolina Corporation Commission, *Annual Report*, 1909, pp. 263–65.
95. AT&T Archives, Raleigh North Carolina Telephone Company, Petition to North Carolina Corporation Commission, n.d. It was no wonder that Southern Bell president W. T. Gentry reported that in North Carolina, "the opposition situation is worse, from a strategic standpoint, than any other state."
96. North Carolina Corporation Commission, *Annual Report*, 1909, In Re Telephone Rates for SBT&T at Fairmount, North Carolina. See also North Carolina Acts, 1899, chap. 164, sec. 2, pts. 10, 11. The act gave the commission powers to make "just and reasonable" rates.
97. North Carolina Corporation Commission, *Annual Report*, 1909, p. 243.
98. Ibid.
99. This did not mean that its rulings always went against Bell. The commission dismissed the charge leveled by an independent promoter that Bell's rates were not "standard" in certain small towns. But its reasons for refusing the petition were that no subscriber in those places had complained. Presumably had local concerns been manifest, the commission would have taken action. North Carolina Corporation Commission, *Annual Report*, 1912, p. 258.
100. AT&T Archives, box 2019, SBT&T Progress Report, 1911, Gentry Report, 1911.
101. AT&T Archives, box 13, Central Union Proposed Reorganiza-

tion, 1910–13, Sunny-Kingsbury, 4 December 1912.

102. House Bills of Illinois, April 1913, bill #573, "An Act in Relation to Public Service Companies." See also House Journal of Illinois, 17 April 1913, pp. 537–64.

103. Section 87 of the 1913 Illinois Public Utilities Commission Law permitted municipalities to own and operate their own utilities if passed by a referendum and subjected to a commission review. *Laws of the State of Illinois Pertaining to State Public Utility Commission* (Springfield, 1914).

104. *Western Electrician*, 1 January 1889, p. 14; 4 May 1889, p. 238. Illinois Senate Bills, February 1889, bill #193, "For an Act to Regulate the Rental of Telephones."

105. Even a dissenting minority report on one public utilities commission bill, while questioning the wisdom of some municipal actions, noted the need to protect home rule and preserve municipal franchise power. House Bills of Illinois, April 1913, bill #664, "Minority Bill from Special Committee." The proposed legislation did not directly favor the interests of cities either. Still, by trying to preserve an active independent telephone industry, the bills would have added to the leverage of city governments in dealing with Bell.

106. AT&T Archives, Illinois, "Report of the Special Committee on Public Utilities," 20 January 1917, p. 6.

107. *Telephony*, 24 May 1913, p. 42; 7 June 1913, p. 35.

108. Illinois Public Utility Commission, *Annual Report*, 1916, pp. 77–78. House Journal of Illinois, 17 April 1913, pp. 537–64. The Journal contains a discussion of the bill. By preventing rates below tariff, the PUC reduced the ability of cities to gain free service; of course this same law was often used to prevent discriminatory practices by carriers against localities. See secs. 36–41 of the 1913 law. *Laws of the State of Illinois*.

109. AT&T Archives, Illinois Bell Management *Annual Reports*, 1911–14; Commercial Department *Annual Report*, 1912. Though relatively large for the nation as a whole, the 25 percent figure indicates nonetheless that even in the Midwest, the region of the fiercest competition, direct confrontations between Bell and independents was simply not the major issue. In Chicago itself, the situation was somewhat different. There, 10,195 of 17,077 subscribers to the local independent firm had two telephones.

110. Illinois Public Utility Commission Cases, docket 7926, 31 July 1918; docket 3756, 9 December 1915.

111. Garnet, *The Telephone Enterprise,* 153–54.
112. Long before the Justice Department had ordered Bell to interconnect with independents, the company had been using a form of interconnection to bring competing companies into its system. The Kingsbury Commitment said nothing about how or on what terms independents had to be granted access to the system. See United States Bureau of the Census, *Historical Statistics,* vol. 2, p. 783.
113. Illinois Public Utility Commission Cases, docket 4462, 4 May 1914; docket 4235, 1916. See also AT&T Archives, box 25, Interstate Telephone and Telegraph Company Acquired by Chicago and Central Union Telephone Company, 1917–18. Also AT&T Archives, box 35, Decatur, Vandalia, and Fayette Companies Acquired by Central Union.
114. In this way, Bell could use sublicense contracts and individual negotiations with municipal governments to integrate more places into its system.
115. Illinois Public Utility Commission Cases, docket 2638, 6 May 1915; docket 4299, 26 June 1916; docket 10362, 18 May 1920. Docket 10362 established that costs incurred in running a line to a customer more than half a mile from an exchange had to be paid by the customer.
116. Illinois Public Utility Commission Cases, docket 7966, 18 June 1918.
117. Some states went even further and banned mutual and cooperative companies. Jones, "Origins of the Certificate," 480.
118. This consensus seems to go back to the early years of telephony in the region (see chaps. 2 and 3).
119. AT&T Archives, Journal of the Manchester Board of Alderman, Manchester, New Hampshire, 5 October 1897.
120. Ibid., 16 November 1897; 1 February 1898.
121. Louis Ashton Thorp, *Manchester of Yesterday* (Manchester, N.H., 1939), 510, 516.
122. AT&T Archives, box 1279, Acquisition of Wells River Valley Company, 1895, Keller-French, 30 September 1895.
123. AT&T Archives, box 1230, Acquisition of Independents in New England, *St. Albans Messenger,* 14 July 1899. AT&T Archives, box 1279, New England Telephone Company Relations–Vermont Telephone and Telegraph, 1897–98, 1900. AT&T Archives, box 1279, Vermont Telephone and Telegraph Company, Organization and Operations, Denver-Keller, 16 August 1897; Note to French,

12 December 1899; *Montpelier Argus and Patriot,* 11 November 1897.

124. AT&T Archives, box 1015, 1911 Convention of New England Sublicensees.

125. Langdale, "The Growth of Long Distance Telephony," 151–52. The South and Midwest experienced growth in their long-distance networks after 1896. At the same time, in these regions large gaps in the system still existed, in contrast to New England. The more limited nature of competition in the Northeast may have allowed Bell more leverage in dealing with municipalities.

126. Massachusetts was an early pioneer in regulation and devised many of the policies which the Midwest and South later followed. In more rural states such as Vermont and New Hampshire, telephones were recognized as a public service industry and subject to common carrier laws before 1894. See New Hampshire Laws, 1881, chap. 54, secs. 11, 12. Journal of the New Hampshire House of Representatives, 1885, p. 387, "Resolution of June 17, 1885."

127. Maine Legislative Record, House, 17 March 1909, p. 765; 30 March 1909, p. 1156; 1 April 1909, p. 1228.

128. Vermont Public Service Commission, Public Service Commission Laws of Vermont, 1911, title 28, chap. 205 of the 1908 law.

129. Twelfth Biennial Report of the Vermont Public Service Commission (hereafter 12 VTPSCR) 1910, pp. 26–29. Vermont Public Acts, 1915, #163, sec. 9.

130. Vermont General Laws, 1917, chap. 212, sec. 5011.

131. New Hampshire Public Service Commission Report, vol. 2 (1912) (hereafter 2 NHPSCR), p. 28. The commission reported to the legislature that it was reluctant to enforce rules prohibiting antidiscrimination. These rules were used elsewhere to restrict the ability of municipalities to extract concessions from telephone carriers.

132. Maine Public Laws, 27 March 1913, #129, "An Act to Create a Public Utility Commission." Maine Public Utility Commission Opinions and Orders, 26 April 1916, "Commission Opinion Requested by Northern Maine Telephone and Telegraph re Authority to Extend Lines." The commission claimed that this was a municipal matter and that it had no jurisdiction over it.

133. Maine Public Laws, 27 March 1913, #129, "An Act to Create a Public Utility Commission," secs. 25, 30, 39, 37. See also Maine Legislative Record, Senate, 18 March 1913, pp. 896–97.

134. Vermont Public Acts, 1908, #116, secs. 15, 17, 23.

135. AT&T Archives, New Hampshire Public Service Commission, "Hearing on Petition of Canaan People's Telephone Company," 17 February 1914.
136. Governor G. H. Prouty Inaugural Address, *Burlington Free Press*, 9 October 1908.
137. 4 VTPSCR, 221–78, "In Re Addison and Panton Telephone and Telegraph."
138. Vermont Public Acts, 1908, #116. Maine Public Laws, 1913, #129, secs., 35, 36, 39.
139. In 1910 the Vermont Public Service Commission recommended eliminating procompetitive elements from the 1909 law. 12 VTPSCR, 26–29.
140. See, e.g., Maine Public Utility Commission, Daily Proceedings, vol. 1, 20 May 1915; 1 June 1915. Maine Public Utility Commission Record, vol. 2 (hereafter 2 MEPUCR), pp. 86–92. Vermont Public Service Commission, "In Re Bethel Telephone Company," 1 June 1915, reprinted in 16 VTPSCR, 412–14.
141. 2 MEPUCR, 102–9, "John F. Goldthwaite v. NET&T," 10 June 1915.
142. 2 MEPUCR, 325–26.
143. 5 MEPUCR, 359.
144. "Citizens of St. Johnsbury vs. NET&T," 20 March 1919, reprinted in 19 VTPSCR, 25–27.
145. Maine Public Utility Commission, *Annual Report*, 1916, pp. 209–15. 18 MEPUCR, 128–29, "Petition of Eastern Telephone and Telegraph to Furnish Certain Service in Washington County."
146. Maine Public Utility Commission Daily Proceedings, vol. 2, 11 July 1916, "Petition of George True and the Maine Telephone and Telegraph Company." See also 6 MEPUCR, 22–23.
147. 3 MEPUCR, 280–82, "Petition of New England Telephone to Sell Telephone Property to Eastern Telephone and Telegraph," 9 April 1910. See also 13 MEPUCR, 386–87, "In Re East Hebron Telephone Company," 8 May 1918.
148. 1 NHPSCR, 111; 2 NHPSCR, 123; 6 NHPSCR, 481.
149. Maine Public Utility Commission, *Annual Report*, 1920, pp. 10–12.
150. AT&T Archives, Illinois, Read Case, vol. 2, 1918, Testimony of F. P. Fish, p. 982.
151. Garnet, *The Telephone Enterprise*, 130–31. AT&T Archives, Illinois, A. P. Allen, Notes on Compulsory Physical Interconnection between Two Competing Companies, November, 1914.

152. AT&T Archives, SBT&T, Florida, Barnett-Carty, 16 June 1911.
153. AT&T Archives, box 1034, SBT&T Incorporation License, 1876–1925, W. T. Gentry, 1909.
154. AT&T Archives, Testimony of O. F. Berry before the Illinois Public Utility Commission, 1919, p. 322.
155. Hughes, *Networks of Power,* shows how municipal interests in London prevented the emergence of a large-scale electrical power transmission system in the region.
156. SBT&T *Annual Report,* 1909. The company feared that it would have to continue to provide service for marginal areas at unremunerative rates.
157. Hughes, *Networks of Power.* See also Christopher Armstrong and H. V. Nelles, *Monopoly's Moment: The Organization and Regulation of Canadian Utilities, 1830–1930* (Philadelphia, 1986).
158. Jones, "Origins of the Certificate," 512–15. Jones notes how regulation can restrict technological change. More generally, it can be used to shape an industry. The Illinois Public Utilities Commission was commended by no less a person than Samuel Insull, the electric utilities magnate who was creating a system of electric power companies, for its work on telephones. AT&T Archives, Illinois, Samuel Insull, "Public Service," May 1919. Richard McCormick notes that state-level progressives took power out of local hands and tended to be more conservative than municipal authorities. Richard McCormick, "The Discovery that Business Corrupts Politics: A Reappraisal of the Origins of Progressivism," *American Historical Review* 86, 2 (1981): 247–74.
159. Midwestern and New England commissions seriously considered ways of reaching rural customers with the system. On the other hand, Illinois did not regulate independent farmers' lines. Many commissions left certain areas beyond their purview because they could not be easily integrated into the system.
160. Georgia Railroad Commission, *Annual Report,* 1914, p. 14.
161. Richard McCormick, "The Party Period and Public Policy: An Exploratory Hypothesis," *Journal of American History* 66, 2 (1979): 279–98. McCormick discusses the distributive approach to public policy, in which the government aided individual localities by providing resources and opportunities for them to exploit, rather than forming a more comprehensive policy that aimed at things such as overall economic efficiency. The distributive approach generally ended with the progressive era, though southerners appear to have stuck to this distributive outlook longer than other Americans.
162. GMLB, 923, 1907, pp. 78–112.

Chapter 9. Regional and Corporate Cultures

1. On this attitude toward big business, see Thomas K. McCraw, *Prophets of Regulation* (Cambridge, Mass., 1984), 80–142.
2. For a general discussion of conflict between northern firms and southern businessmen, see C. Vann Woodward, *Origins of the New South* (Baton Rouge, La.,1951), 142–74, 264–90.
3. Wright, *Old South, New South,* 33–34. Wright emphasizes the impact of the end of slavery on the location of economic activity.
4. Contrast Jonathan Wiener, *The Social Origins of the New South: Alabama, 1860–80* (Baton Rouge, La., 1978). Also, Jonathan Wiener, Robert Higgs, and Harold Woodman, "Class Structure and Economic Development in the American South, 1865–1955," *American Historical Review* 84, 4 (1979): 970–1006. Dwight Billings, *Planters and the Making of a "New South": Class, Politics and Development in North Carolina, 1865–1900* (Chapel Hill, N.C., 1979), argues that southern businessmen embraced industry only to support paternalistic social relations.
5. Contrast Robert Higgs, *The Transformation of the American Economy, 1865–1914* (New York, 1971).
6. Chandler, *The Visible Hand,* 148–88, describes how railroad system builders united with large financiers to override opposition from small, local businessmen. Bell did the same thing, but it made key compromises that allowed local capital a role, although a subordinate one.
7. See Robert Wiebe, *The Segmented Society* (New York, 1975), and Thomas Bender, *Community and Social Change in America* (New Brunswick, N.J., 1978).
8. Government regulators also enacted some compromise policies, but the final result conformed more to the needs of the Bell system than to those of southern consumers. Schumpeter, *The Theory of Economic Development,* suggests that monopoly aids innovation. More generally, the control of key technologies, scarce resources, or the mobilization of political power may be a necessary part of the innovative process, particularly when innovators run into opposition.
9. Carlton, *Mill and Town in South Carolina,* 40–82, discusses urban developers and industrialists in the New South. See also Rabinowitz, "Continuity and Change in Southern Urban Development."
10. By the 1920s, further technological advances were also lowering the cost of long-distance and local service, erasing the price advantage enjoyed by independents.
11. One could make assumptions about the value of time to consumers,

as was done in the case of business users, and then, theoretically at least, calculate the social savings made possible by telephony and compare it to the investment in the telephone system. Assumptions about the value of time, particularly to individual consumers, are highly speculative, however. And all of this would still ignore the qualitative value of voice communication as compared to other forms of communication.

12. For examples of how those alternatives might work, see Joel Tarr et al., "The City and the Telegraph: Urban Telecommunications in the Pre-Telephone Era," *Journal of Urban History* 14, 1 (1987): 38–80.
13. Peter Temin, with Louis Galambos, *The Fall of the Bell System* (New York, 1987).
14. Fisher, "Technology's Retreat." By 1920 over 65 percent of the farms of the Midwest had telephones; nationwide only 38.7 percent did, and in the South, only 11.5 percent. United States Bureau of the Census, *Census of Agriculture* (Washington, D.C., 1952), vol. 2, p. 211. Those numbers declined after 1920, due in part it seems to AT&T policy.
15. Armstrong and Nelles, *Monopoly's Moment,* 82. The authors note this problem in the case of Canadian cities.
16. A good example of such feasible adjustments are those which the company made in the early 1900s to connect more farmers to the system (see chap. 7).
17. William Parker, "The South in the National Economy, 1865–1970," *Southern Economic Journal* 46, 4 (1980): 1044–45. Parker notes how marginal areas are frequently ignored by outside capitalists, which slows down the process of resource transfer. Northern dominance of the southern steel industry may be a case where the South was hurt by contact with outside firms. Wright, *Old South, New South,* 165–71, does not believe so, however.
18. Wright, *Old South, New South,* 14–15.
19. Witness Southern Bell's ability to train and install southerners in managerial positions over time. It is true that acquiescence meant that southerners' influence on the corporate world would have been confined to marginal adjustments. But these adjustments were not necessarily insignificant. The only other option was renouncing the modern economy for some alternative one.
20. Robert Dykstra, "Town-Country Conflict: A Hidden Dimension of American Social History," *Agricultural History* 38, 4 (1964): 196–99. Dykstra discusses this problem in the West.

Index

References to pictures and tables are in italics.